Fizzics

Fizzics

The Science of Bubbles, Droplets, and Foams

F. RONALD YOUNG

THE JOHNS HOPKINS UNIVERSITY PRESS
Baltimore

The Johns Hopkins University Press
2715 North Charles Street
Baltimore, Maryland 21218-4363
www.press.jhu.edu

Library of Congress Cataloging-in-Publication Data

Young, F. Ronald.
Fizzics : the science of bubbles, droplets, and foams / F. Ronald Young.
p. cm.
Includes bibliographical references and index.
ISBN-13: 978-0-8018-9891-4 (hardcover)
ISBN-10: 0-8018-9891-9 (hardcover)
ISBN-13: 978-0-8018-9892-1 (pbk.)
ISBN-10: 0-8018-9892-7 (pbk.)
1. Gases—Popular works.   2. Bubbles—Popular works.   I. Title.

QC161.Y68 2011
530.4'275—dc22   2010045577

A catalog record for this book is available from the British Library.

Special discounts are available for bulk purchases of this book.
For more information, please contact Special Sales at 410-516-6936 or
specialsales@press.jhu.edu.

The Johns Hopkins University Press uses environmentally friendly book
materials, including recycled text paper that is composed of at least
30 percent post-consumer waste, whenever possible. All of our book
papers are acid-free, and our jackets and covers are printed on paper
with recycled content.

To my grandchildren:
Nicholas, Antonia, and Joseph

The flow of the river is ceaseless and its water is never the same.
Bubbles that float in the pools break and vanish while others form to
replace them. None of them lasts long. In this world, people and their
dwellings are like that, forever changing.

—*Kamo no Chōmei (1155–1216), Hōjōki*

For many years I taught physics at Watford College, which is just a few
miles north of the hustle and bustle of central London. Teaching has always been my chief love, but I was fortunate enough to be able to set up a
research laboratory at the college, a place where we could explore cavitation. Over the years I've lost track of the number of times I've told
people—including fellow scientists—that I investigate cavitation, only to
see them look puzzled and ask, "What exactly *is* cavitation?" To answer
that question, taking into account the complex underlying physics and
mathematics, I wrote a book on the subject, back in 1989. (Summed up
briefly, cavitation is the spontaneous formation of bubbles in a liquid.)

In more recent years, my attention has been captured by a related—
and bizarre—phenomenon known as sonoluminescence. When a bubble in a liquid bursts, it can, under some circumstances, generate not
only sound but an eerie blue light as well. This baffled many clever scientists, but eventually we came to understand what causes it. Again I
wrote a book, aimed at researchers, to explain what we know about it.

Now, though, I want to tell a different story. Although cavitation and
sonoluminescence are parts of that story, they are but small ones. I want
to travel back to my childhood and remember how thoroughly I enjoyed

blowing bubbles! Later I watched my children blow bubbles, still later my grandchildren. Bubble blowing joins the generations together—and, notwithstanding television, computers, the Internet, and all manner of other marvels still to be invented—it probably always will. I want my grandchildren to grow up feeling some of the delight, and mystery, that science can hold. To help them along the way, I thought I'd write a book about bubbles and their closely related cousins, the droplets and foams—a book that many people can easily understand, a book that presents no mathematics or complex formulas but rather tries to share the *joy* of bubbles, both in nature and in the laboratory. One of the few formulas in the book is a recipe for a soap mixture that produces big, long-lasting bubbles—I hope you will try it.

This book is brief, and each chapter is quite short. The book's brevity, combined with its copious illustrations, should make it understandable to most everyone, including readers with little or no scientific or mathematical knowledge.

The prologue offers a rapid introduction to the five different types of "bubble," as distinguished by centuries of study. Chapter 1 then describes air bubbles found in water, and chapter 2 deals with their opposite—water droplets found in air. Chapter 3 explains foams, ranging from the head on a glass of beer to the fire-fighting foam used to smother the engines of crashed aircraft to prevent flames from spreading beyond.

Chapter 4 looks at soap bubbles and soap films, explaining, for instance, how some insects are able to walk on water. I also show you how to make a soapy computer. Among other things, chapter 5 describes how humpback whales use bubbles to cast nets and catch fish.

Chapter 6 explores my former field of research, sonoluminescence—the bluish-white light sometimes produced when bubbles in liquids collapse. Although humans can create this phenomenon in a laboratory, we'll see how the snapping shrimp uses it to knock unconscious the other small marine creatures that it wants to eat.

The final chapter shows how the humble bubble plays a literally vital role in medicine. Here we look at decompression sickness, bubbles—

good and bad—in the human blood system, and how bubbles can be used to treat kidney stones or clean teeth.

Books are rarely written single-handedly. For help in producing this book, I would like to thank Robert Young, Michael Postema, Timothy Leighton, Ann Pantling, Paul Evans, Christine Foucault, Grant Romain, Neil Ferguson, and many others. I am much indebted to Anne Hartley for typing the manuscript with speed and great accuracy and to Greg Nicholl of the Johns Hopkins University Press for drawing the admirably clear diagrams.

A special tribute of thanks is due to Professor Detlef Lohse and Dr. Cyril Isenberg, who read the manuscript and made many informative suggestions. Dr. Isenberg, whose own book on bubbles stands as a model for us all, also kindly supplied a CD containing reproductions of paintings involving bubbles, which proved to be useful. Professor Denis Weaire, a leading authority on foams, commented helpfully on chapter 3, which led to Professor Jan Cilliers supplying me with authoritative information on the processing of minerals.

Finally, I owe an enormous debt of gratitude to Trevor Lipscombe, my editor at the Johns Hopkins University Press. Without his generous help at every stage—and particularly during the past year—this book would never have seen the light of day.

Prologue

I'm forever blowing bubbles,
Pretty bubbles in the air.
They fly so high,
Nearly reach the sky,
Then like my dreams, they fade and die.
Fortune's always hiding,
I've looked everywhere,
I'm forever blowing bubbles,
Pretty bubbles in the air.

—Joan Kenbrovin
and John William Kellete,
"I'm Forever Blowing Bubbles"

A bubble has the perfect shape—at least according to the Ancient Greeks. Aristotle, tutor to the warrior king Alexander the Great, spent a few pages of his book *On the Heavens* discussing the superiority and perfection of the sphere, the shape he used to construct his model of the universe. And in an ideal universe, a bubble is a simple sphere. It has no corners, no edges, and just one number, its radius, which completely describes it. And this beautiful, spherical shape is stable. A bubble *wants* to be a bubble.

Bubbles are part of everyday life. They whiz and fizz in a glass of soda, and a bubble bath soothes away the aches of a working day. But bubbles can bring life or death: a large bubble trapped in a blood vessel, an embolism, can be fatal. A diver who surfaces too swiftly after a deep dive may be in for trouble. Any excess nitrogen dissolved in the diver's blood (the air we breathe is four-fifths nitrogen) may start to form bubbles in the bloodstream, producing the agonizing pains known as "the bends."

Scientists have come to the view that there are five different types of bubbles. The first, familiar to us all, is an air bubble in water (Fig. P.1a), which can range in diameter from a hundredth of a millimeter to several millimeters. An air bubble in water is, more generally, a bubble of gas in a liquid, just like the bubble of nitrogen in the blood. The second type is not really a bubble, but—technically speaking—a droplet, such as a water drop in air (Fig. P.1b). It's the opposite of the air bubble in water, in the sense that it forms where water replaces air, rather than where air replaces water. This, too, is merely a case in point: a sphere of some other liquid immersed in some other gas is also a droplet.

Another form of bubble is found in foam (Fig. P.1c), as a gas bubble only partially immersed in liquid. A fine example is the foam that forms on a freshly poured pint of beer (Fig. P.2), known colloquially as the head. And, at least in England, distinctions concerning the head are important: in the North of England, beer is served with much froth and foam atop the surface, whereas in the South, it's more common to serve beers with little or no head. Southerners, in fact, enjoy "a dead pint," which is a pint of pure liquid, without much foam at all.

We also have the humble, lovable soap bubble (Fig. P.1d), which is more complex than it first appears. Technically speaking, these bubbles

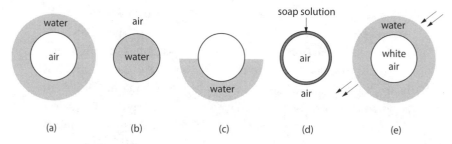

Figure P.1. Five different types of bubbles: (a) an air bubble in water, (b) a water drop in air, (c) a foam gas bubble partially immersed in a liquid, (d) a soap bubble, and (e) an air bubble trapped in moving water. (Greg Nicholl)

Figure P.2. Foam (or "head") on a freshly poured pint of beer. (Trexer/ Wikimedia)

are films, for a thin layer of liquid (soap) is sandwiched between inside and outside layers of gas (air). We can leave it to scientists to debate whether bubble gum falls into this category: experts produce a thin film of bubble gum with air on either side, but whether chewed gum is solid or liquid is an open question. Closely related are helium balloons—ever present at children's parties—and hot-air balloons: both have a gas (perhaps air) on the inside, and air on the outside, with a thin layer of some material or other between.

The final animal in the bubble zoo is an air bubble trapped in moving water. These "entrained" bubbles, formed and reformed by the millions, make the water appear white—as those who go white-water rafting or kayaking know full well—and such bubbles generally form in waterfalls as well (Fig. P.1e).

CHAPTER ONE

The Physics of Fizzy

Gas Bubbles in Liquid

Double, double, toil and trouble;
Fire burn and cauldron bubble.
—*Shakespeare, "Macbeth"*

Bubbles consisting of gas in a liquid occur widely. Air bubbles in water, the most familiar case in point, range from the bubbles you find in a glass of water to those found in the sea, both of which are formed from air dissolved in water. We'll have a look here at a dozen-odd circumstances where bubbles play different roles.

Bubbles in Beer and Volcanoes

One of the great experiments in physics is to examine the bubbles that appear in freshly poured beer. Streams of bubbles rise from particular places on the surface of the glass (Fig. 1.1). No matter how smooth the glass may seem, its inner surface has tiny cavities or cracks, which trap minuscule bubbles of carbon dioxide gas. Why the carbon dioxide? Because the major breweries pump large amounts of it into the beer and then seal the bottles under high pressure. Open a bottle from one of Dublin's major brewers and you instantly reduce the air pressure above the beer. The carbon dioxide bubble that gripped the micro-crack in the glass so tightly is now under reduced pressure, and starts to expand. At some point, it reaches a size where it breaks free from the crack and rises as

Figure 1.1. Bubbles rising from particular places on the inner surface of a glass. The glass on the left has a "widget" etched into its base that causes more bubbles to form there. (Wikimedia)

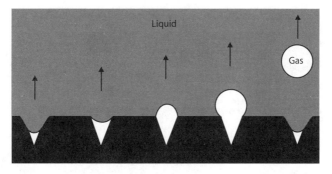

Figure 1.2. Creation of a carbon dioxide bubble that is in due course released from a micro-crack in glass. (Greg Nicholl)

a separate bubble (Fig. 1.2). The pressure in a liquid or a gas increases with depth (that's why your ears pop when your airplane comes in to land), and so, as the carbon dioxide bubble rises through the Guinness, the pressure in the beer surrounding it diminishes and the bubble expands. The net result from pumping all that extra carbon dioxide into the beer is an ample foam—a rich pile of bubbles—atop your glass. Enjoy!

By the way, the famous scientist James Prescott Joule (1818–1889), after whom the metric unit of energy is named, came from a long line of brewers. Joule's bitter beer is still available in the North of England, though, sad to say, it's been taken over by a major brewery. Joule's father, Benjamin, built James a laboratory in his house in Sale, Cheshire. James also had the good fortune to be taught science by John Dalton, one of the major contributors to the modern atomic theory of matter, and went on to devote forty years to exploring the relation between work and heat. He made many contributions to physics, but arguably his most important is the first law of thermodynamics. Close to the town of Sale is Manchester, with its huge Victorian Town Hall. Inside its front door are two statues, one of which is John Dalton (on the left), the other James Joule (on the right).

But let's move on. Bubbles of air in water will rise, simply because they are lighter than water. This is something we all know, but engineers have put it to good use for centuries in a humble but amazing device: the spirit level. The ordinary spirit level, often called simply a "level," consists of a sealed, slightly curved glass tube set in a wooden or metal casing. The tube is filled with colored liquid, but not quite completely, so that a bubble of air remains. This bubble will rest at the dead center of the tube only if the surface on which the level rests is perfectly horizontal. If you ever want to upgrade things around the house, such as put in a new door or kitchen cabinet, the spirit level—a bubble in a tube—will show you just how off-kilter your walls may be.

Sometimes, though, simple gas bubbles can be devastating. In a standard science fair experiment, you make a cone-shaped cardboard container and fill it with baking soda. Add some vinegar and massive amounts of carbon dioxide are released, causing foam to push out of the cone and down its sides. This spectacular effect is a model for the behavior of a real-life volcano. Down within a volcano's depths is magma, and dissolved within this exceedingly hot liquid are water vapor, carbon dioxide, sulfur dioxide, and other gases. But if the magma cools, or is decompressed, these gases form bubbles, and the bubbly magma, now so much greater in volume, will spew up out of the volcano, possibly causing great devastation.

The Ins and Outs of Bubble Oscillations

The ideal bubble is perfectly spherical and has a definite diameter, but things can change. Scientists like to ponder what might happen if something simple is tweaked a bit. What happens, for example, if the pressure of the air around a droplet of water is altered ever so slightly? The answer is that the bubble oscillates. Its radius will grow greater, then smaller, oscillating in this fashion about the untweaked, original diameter. What could cause air pressure to change? Sound can.

Sound is something we take for granted. To a scientist, sound is caused by vibrations in the air around us. That is to say, tiny changes in the density of the air and in air pressure spread rapidly through the air. At the end of its journey, this wave of density and pressure change can be recorded by an extremely delicate detection system: the ear of humans and so many other animals. Sounds in water originate from the bubbles present in the sea, in rivers, and in the pipe flows in our houses and buildings. If the radius of a bubble immersed in water oscillates back and forth, it produces corresponding changes in the density and pressure of the water similar to those produced in air. Rainfall, in contrast, produces sound in two ways: first, when a droplet strikes the solid surface at ground level and bursts, and, second, when a droplet becomes a bubble in the water that has begun to collect. The collision, in either case, causes the bubble to vibrate radially, producing sound. In 1933, the Belgian astronomer Marcel Minnaert showed that the almost musical sound of running water is caused by countless air bubbles oscillating at their natural frequencies.

Gas bubbles in liquids can also produce sound when they oscillate, which has several practical consequences, such as gargling. Professor Werner Lauterborn at the University of Göttingen in Germany, to study gas bubbles, focused pulses of laser light onto silicone oil. The laser light causes the bubbles that exist in the oil to grow to a diameter of a few millimeters, but then, when the pulse is off, they collapse to their initial size. (Why the laser light causes this growth is a complicated issue, worthy of its own research article.) Professor Lauterborn's bubbles were photographed by a high-speed camera, which took photos at the phenomenal rate of 75,000 frames per second. Figure 1.3 shows a sequence of fifty

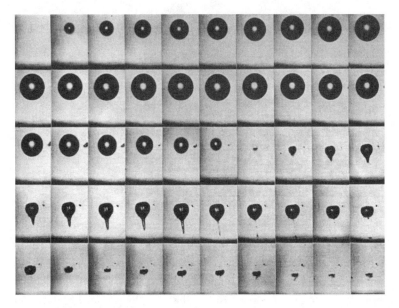

Figure 1.3. A sequence of photographs of a bubble near a solid wall. These fifty illustrations (read from top left to bottom right) show the change in size and shape of the bubble over time. (Butterworth and Co.)

photographs of a bubble near a solid wall. The bubble grows to a maximum size and decays to a minimum three times. To give a sense of scale, the bubble was produced at a distance of 5 millimeters from the wall and reached a maximum radius of 2 millimeters. The photographs show how complicated bubble oscillations can be!

Springboard and Platform Diving

Every four years the Olympics come around. One event, first made popular by the incomparable Greg Louganis, is diving. From a springboard or a high platform, competitors arc gracefully through the air and into the water below, trying to make the smallest splash possible. But because people are difficult to study, scientists studying the entry of objects into water prefer to work with things such as a cylindrical body with a rounded nose. What scientists learn from such a simple object, they hope will apply to a more complex one. When a round-nosed, cylindrical body enters the water, a cavity forms behind it. This pocket lengthens, but eventually closes in. Springboard divers know all this full well: they deliberately use their hands to create a "pre-splash" (Fig. 1.4), which produces a small cavity through which the diver's body moves an instant later. By doing this, divers minimize splash (Fig. 1.5) and maximize their score.

Propellers, Pumps, and Porpoises

The S.S. *Titanic*, arguably the most famous ship ever built, had huge propellers. The left and right propellers, each three-bladed, weighed in at 38 tons apiece and measured 23' 6" in diameter. The four-bladed central propeller was a mere 17' 0" and a lightweight 22 tons, but could rotate at 165 rpm. The largest ship propeller produced thus far, cast by the Hyundai Corporation, is a 101-ton monster with a diameter just shy of 30 feet.

Figure 1.4. A diver using her hands to create a "pre-splash." (U.S. Navy)

Figure 1.5. Result of the "pre-splash," which produces a small cavity through which the diver's body then moves. (Bob Thomas/iStockphoto)

But quickly rotating propellers, large and small, face a problem. When a ship is under way and the propellers rotate swiftly, the blades sweep out a helical pattern of bubbles suggesting a corkscrew (Fig. 1.6). The bubbles, though, can be a problem, for they are caused by *cavitation*. As the blades rotate, they increase the speed of the water close to the blade. The higher the speed of a fluid, the lower its pressure becomes. If the speed becomes too great, and the pressure too low, vapor bubbles are formed in the liquid. But the vapor bubbles are unstable and can collapse, generating high-pressure waves that can damage the propeller. Their collapse also causes noise, which is a major difficulty for warships that wish to remain undetected by enemy vessels—submarines, for example.

This process of bubble formation, called cavitation, was first predicted by the Irish scientist Osborne Reynolds, long before it was observed in the trials of HMS *Daring* in 1893. HMS *Daring*, a destroyer armed with torpedos, was known at the time as the fastest boat ever, logging better than 28 knots. This, however, was below its expected speed, and engineers attributed this disappointing performance to the formation of water vapor bubbles on the propeller blades. A wider propeller blade was soon

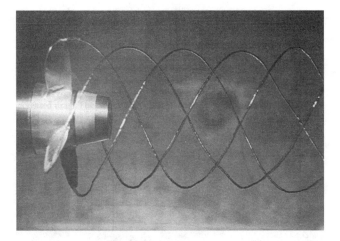

Figure 1.6. A helical pattern of cavitation bubbles created from blades rotating in water. (Garfield Thomas Water Tunnel, Pennsylvania State University)

designed to eliminate the problem, but HMS *Daring*'s days were numbered, and the solution was set aside. In 1897, HMS *Turbinia* was commissioned, the first vessel of His Majesty's Royal Navy to be powered by a steam turbine. The steam-turbine vessels had come to stay, but their higher cruising speeds meant that cavitation issues were about to become a widespread problem for the fleet, for decades to come.

Cavitation severely erodes propellers (Fig. 1.7). As the blades sweep around, they create areas of low pressure within the water, by sucking air from the atmosphere. Beyond a certain speed, these cavities are formed close to the blade's surface, and when the cavities collapse they send shock waves through the water that in time can damage the propeller. The cavitation region usually appears as a cloud of bubbles of different sizes, which collapse at different times, producing considerable noise. To avoid this problem, engineers often try to coat propellers in a nonmetallic substance, which absorbs the shock waves better. The problems persist today.

Around the home, there are plenty of pumps to be found. The hot-water system relies on one, as does a washing machine. If you're lucky enough to have a dishwasher, that relies on a pump too, and so does your

Figure 1.7. Erosion of propeller blades due to cavitation. (Erik Axdahl/ Wikimedia)

Jacuzzi. The blades in the pumps, though, are also afflicted by cavitation damage, and to avoid it, engineers design pumps with an increased inlet diameter, increased pressure, and a reduction in the speed of the fluid moving through the pump. The result is a quieter home, and one where pumps don't have to be replaced because of cavitation erosion.

Animals can feel the effects of cavitation, too. Dolphins, for example, are highly streamlined, powerful swimmers, but they also have a well-developed nervous system. So, when a dolphin swims too swiftly, the more violent flicking of its tail can cause cavitation to set in, and the results will be painful. Just like HMS *Daring*, dolphins could probably move more rapidly through the water were it not for cavitation.

Knuckle Cracking

Etiquette authorities, such as Miss Manners, warn vigorously against cracking your knuckles in public. Scientists, though, seldom read the words of Miss Manners, and instead regard the cracking of the metacarpophalangeal joint in one or more knuckles as another example of the effects of bubbles. Figure 1.8 shows a cross section through the metacarpophalangeal joint. It is basically a ball-and-socket joint filled with synovial fluid. Expert knuckle crackers force their fingers into strange positions, something gives, and there is a sudden increase in bone separation. This increase in separation causes vapor and gas bubbles in the synovial fluid to be set free from the liquid. It is the rapid collapse of the bubbles that causes the audible crack.

A detailed study of joint cracking was carried out at Leeds University in England. Researchers devised a machine to look at the force (technically, the load) exerted on the metacarpophalangeal joint of the middle finger, as well as the separation between the bones meeting at that joint. Of the seventeen subjects in the study, five produced cracks, seven did not, and the others were not relaxed enough to enable proper tests to be performed. The scientists also created a model of the metacarpophalangeal joint, using nylon for the metacarpal head and Perspex (Plexiglas or Lucite) for the proximal phalanx. They injected some synovial fluid into this artificial joint and then compressed it. When the artificial joint was suddenly pulled apart, it produced an audible crack. Successive frames from high-speed film showed the formation and collapse of bubbles in the model.

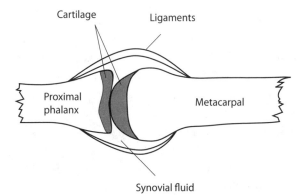

Figure 1.8. Cross section through the joint of the middle finger. (Greg Nicholl)

No study of joint cracking would be complete without a look at its biological implications. A journal popular with scientists, the *Annals of Improbable Research*, hosts the Ig Nobel Prizes each year, to honor Alfred's little-known brother Ignatius (rest assured, he doesn't exist). The prizes honor some of the strangest projects carried out in the name of science, and are always dispensed with good spirit and great humor. In 2009, the Ig Nobel Prize for Medicine went to Donald Ungar, who had cracked the knuckles of his left hand every day for fifty years—but had never cracked the knuckles of his right hand. He was honored for his article "Does Knuckle Cracking Lead to Arthritis of the Fingers?" which was published in the scholarly journal *Arthritis and Rheumatism*, vol. 41, no. 5, 1998, pp. 949–50. Knuckle crackers rest easy: there was no difference between his two hands after all those years, cavitation or not!

Dams

The Tarbela Dam in Pakistan spans the mighty Indus River. The Tarbela, first filled in 1974, is one of the world's largest dams. It converts the energy of the water sweeping down from the Himalayas into great quantities of relatively cheap and environmentally clean hydroelectric power. But even here, cavitation reared its ugly head. Four tunnels, deep within the natural rock around the sides of the dam, controlled the rate at which the reservoir level rose and fell. Each tunnel incorporated three gauge gates—each gate 4.5 feet high and 4.5 feet wide—blocking a tunnel 45 feet in diameter. As the dam filled, one of these gates jammed half-open. Engineers later discovered that with that central gate ajar but with both side gates closed, intense cavitation was generated. The cavities, or bubbles, that formed on the gates eventually collapsed farther down the tunnel, severely eroding the concrete lining. The damage worsened until a hole eroded through the 6-foot-thick reinforced lining of the tunnel. Disaster was at hand: a vast cavern opened within the mountain, one that would collapse the tunnel (Fig. 1.9). Millions of gallons of water (a pint of water weighs a pound) shot out of the tunnel every second in a 100-mile-per-hour jet, bearing with it enormous boulders from the mountainside and chunks of concrete from the tunnel. The only thing left to do

Figure 1.9. Collapse of the tunnel beneath the Tarbela Dam, in Pakistan. (F. Ronald Young)

was to empty the reservoir and reconstruct vast parts of the dam and its environs.

Cartesian Divers

Even toys involve bubbles. One example is the Cartesian diver (Fig. 1.10), which consists of a hollow figure with a small opening in its bottom. The figure is encased in a tall jar or plastic bottle partially filled with water. This container is covered tightly with a rubber membrane. Push the membrane inward and you generate extra pressure, which is transmitted through the water to the diver. Water enters the diver, compressing the air it contains. This process increases the mass of the diver, making it sink. When you release the membrane, everything is reversed: water flows out of the diver, and it rises. If, instead of a glass jar with a membrane, you have a clear, closed, plastic bottle, simply squeezing the sides of the bottle makes the diver submerge.

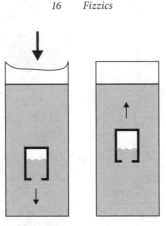

Figure 1.10. Cartesian diver. Push down on the membrane at the top and the diver sinks (*left*). Release the membrane, lowering the pressure, and the diver rises (*right*). (Greg Nicholl)

Bubbles in Your Bubbly

What wedding would be complete without a champagne toast to the happy couple? The cork flies, the champagne gushes out, glasses are poured, and everyone takes a quick sip to honor the bride and groom. The bubbles in the bubbly, as with beer, are filled with carbon dioxide gas released from tiny cracks in the glass. If enough gas accumulates at one of these "nucleation points," it forms a bubble. This tiny bubble grows until it breaks away and floats up to the surface, leaving the nucleation point free for another bubble to form. The result is a constant stream of bubbles that seem to arise from nowhere (Fig. 1.11). The bubbles are released with clockwork regularity. Because they are less dense than the liquid, they begin to rise, just as a helium balloon rises because it is less dense than the surrounding air. But something else happens to a champagne bubble. The pressure in a fluid increases with depth, which is why your ears pop if you dive to the bottom in a deep swimming pool or when your plane lands (you're going deeper into the atmosphere). So, as the bubble rises, the pressure the liquid exerts on it decreases, and the bubble expands. This makes the bubble less dense, and as a consequence it rises faster. That's what you see in the figure: the bubbles get bigger as they rise and get farther apart as they accelerate upward.

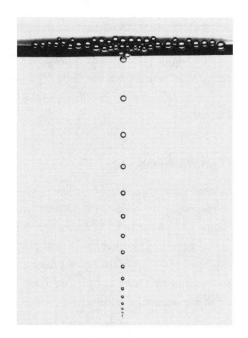

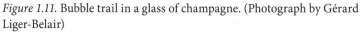

Figure 1.11. Bubble trail in a glass of champagne. (Photograph by Gérard Liger-Belair)

There is a reason for the shape of the classic fluted champagne glass. It has a long stem (so you can hold your glass comfortably and heat from your hand won't warm the wine), but the glass's narrow opening, with its small surface area, makes it more difficult for the crowded bubbles to leave the liquid and enter the atmosphere. Instead, many of them stay in place as a lovely fizzy foam at the top of the glass.

Champagne, a sparkling wine, is made by fermenting grape juice in such a way that the sugars in the juice are gradually converted by yeast into alcohol and carbon dioxide. Ordinary wine is bottled after fermentation is complete, but with champagne, the last stage of fermentation happens inside the bottle. That way, more carbon dioxide is trapped inside and gas pressure builds up. That's why champagne bottles must be thicker than wine bottles, and why the cork is held in place with wire. When the bottle is uncorked, the gas pressure is released with a loud pop. Moments later, bubbles of gas released low in the champagne may push some of the fizzy liquid out of the bottle—especially if the bottle's been shaken and the bottle's opening is partially blocked by a finger or thumb! The winners of motor racing competitions enjoy doing this! The same thing can happen with soft drinks, but for these drinks the carbon dioxide is pumped into the liquid under pressure instead of being generated through fermentation.

The Ascent of Sap in Trees

If you like pancakes or Belgian waffles, you probably know about sap. In winter time, the sugar maple tree stores sugar deep within its roots. Come spring, the tree pumps sap upward, to bring water from its roots to its budding leaves. The good people of Quebec, the biggest producers of maple syrup in the world, know this and tap the sap to produce the sweet sticky syrup we love. But the atmosphere can support a column of water no more than 10.4 meters high, and some trees are much taller than this. The imposing giant redwoods of California can attain heights of more than 100 meters. So how can the sap rise to the top of a giant redwood? The answer is that there can be negative pressure in the column of sap. When water evaporates through pores in the leaves, negative pressure is apparently created in the thin capillaries that bring water to the leaves. It is this negative pressure from above, rather than a positive pressure from below, that causes the sap to rise. In other words, the sap is not pushed up from the bottom, but rather sucked up from the top, in the fashion of a giant drinking straw.

To demonstrate that negative pressures can exist, botanists use a so-called pressure bomb (a.k.a. a Scholander bomb). This is a sealed jar in which a leaf is placed such that its stem is pointing upward and poking out of the jar's stopper. If the leaf cells are truly causing negative pressure, then sap will be sucked up by the leaf, leaving the stem sap-free. The next step is to use a compressor to increase the air pressure in the jar, thereby squeezing the leaf. Eventually, the pressure inside the jar counteracts the negative pressure generated by the leaf, which makes sap appear on the stem. The botanist then knows that the air pressure inside the jar is the same as the negative pressure generated by the leaf.

This simple experiment suggests that negative pressures truly exist, and that they range from about minus 4 atmospheres in damp forest trees, to minus 80 atmospheres in some desert plants, which strive to suck up every drop of water from their dry environment.

The vessels that transport sap from the roots to the top of a tree are not always filled with one continuous column of water. Sometimes, when the negative pressure exceeds a certain value, the continuous columns break, bringing on cavitation—in the form of cavities within the air and vapor. In one experiment, a sensitive microphone acoustically detected the onset of this cavitation. Before cavitation ensues, the walls of the sap vessels are pulled inward by tension in the sap column. When the column breaks, the walls relax and vibrate. When amplified, these vibrations are heard as clicks. More about this remarkable finding can be found in the book *Plants and Water* by J. F. Sutcliffe (New York: St. Martin's Press, 1979).

Ceramic Bubbles

Potters create artistic ceramic designs using plaster molds. These molds are filled with what is called casting slip, a liquid mixture that potters, or industrial potteries, must blend together to form a smooth texture. Mix too swiftly, and bubbles may form in the slip. Then, when the mixture is fired in the kiln at temperatures of around 2000°F these bubbles will expand and ruin the vase (or whatever other product) that's being manufactured. To prevent such a problem, an acoustic device was recently invented. This device sends sound waves into the molten mixture, and the echos the casting slip sends back will be markedly different if bubbles are present, rather like the way sonar detects submarines. The procedure allows pottery companies to avoid producing damaged goods, even though they can't see inside the kiln, and even though the slip is completely opaque.

White-Water Physics

Before going I took a last look at the breakers, wanting to make out how the comb is morselled so fine into string and tassel, as I have latterly noticed it to be. I saw big smooth flinty waves, carved and scuppled in shallow grooves, much swelling when the wind freshened, burst on the rocky spurs of the cliff at the little cove and break into bushes of foam. In an enclosure of rocks the peaks of the water romped and wandered and a

light crown of tufty scum standing high on the surface kept slowly turning round: chips of it blew off and gadded about without weight in the air.
　　　　　　　—Gerard Manley Hopkins, journal entry, August 16, 1873

Anyone who has floated down a mighty river, whether on a large raft or in a single-person kayak, knows firsthand what white water looks like. This phenomenon occurs not only in turbulent rivers, but also in waterfalls and in the waves that break at the seaside. Other examples are the water jets for the propulsion of ships, and—if you partially block the opening with your fingers—white water can be made to stream from garden hoses. You can even produce white water from a faucet, especially in cold weather: many bubbles of air are trapped in the water, and the light that strikes the surface of the bubbles is randomly scattered. This makes the water in your glass appear white. As water goes over the Victoria Falls or Niagara Falls, plenty of air is mixed in with the water, yielding the same white appearance. The same holds true for fast-flowing rivers, particularly around the boulders so loved by experienced kayakers (Fig. 1.12).

Figure 1.12. Fast-flowing rivers, so loved by kayakers, also appear white, owing to the reflection of light by the air that is mixed in with the water. (Photos.com)

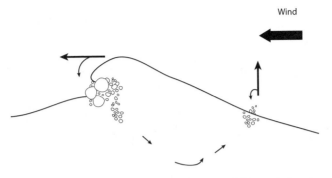

Figure 1.13. A breaker in the surf eventually crests, folding air into its surface layer, which then forms whitecaps on the wave's forward slope. (Greg Nicholl)

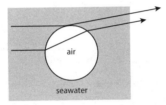

Figure 1.14. Light reflected and refracted through a bubble in the crest of a wave. (Greg Nicholl)

The technical term for air bubbles trapped in water is *air entrainment*. It's a complicated process, and the explanation offered below, as to why whitecaps in the ocean are white, is greatly simplified. As "spilling breakers" come close to the shore, they steepen. These are the kind of waves you get on gently sloping beaches, particularly when the wind blows from land to sea. A spilling breaker will eventually form a sharp crest or peak from which water then breaks, folding air into the wave's surface layer, and forming fairly steady whitecaps on the wave's forward slope (Fig. 1.13). Light from the sky is reflected and refracted by some of these small bubbles, and reaches the observer as shown in Figure 1.14.

For whitecaps to form, the sea water must have a thin film of organic matter on its surface, but because the sea is not pure water but contains seaweed, salts, and oils, this will always be the case.

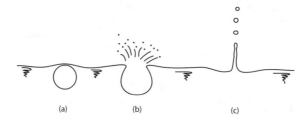

Figure 1.15. Oceanic bubbles reaching the surface (a), which produces two types of droplets: (b) the film drop and (c) the jet drop. (Greg Nicholl)

As the wave breaks gently at the crest, it traps enough air bubbles (as air entrainment) for the resulting air-water mixture to be significantly lighter in weight than the water below it. This density difference will inhibit mixing water from beneath with the surface of the wave, so that the whitecap continues to ride atop the sloping sea surface. When some of these oceanic air bubbles reach the sea surface (Fig. 1.15a), they burst, producing two families of droplets, the film drop (Fig. 1.15b) and the jet drop (Fig. 1.15c).

These two families of droplets are created by different processes. As the ocean bubble rises close to the surface, there's just a thin layer of water separating the gas in the bubble from the air in the atmosphere. Eventually, this layer ruptures, launching film drops from the ocean surface that reach a height of about 1 centimeter. But the breaking of that film leaves a cauldron of gas at the surface, and this collapses dramatically, sending a jet of water high into the air when the surrounding ocean bubbles fill the void. This vertical jet of water soon breaks up, producing somewhere between one and ten jet drops, each of them about one-tenth the size of the ocean bubble that created them.

Jet drops are beautiful to behold, but there is a practical consequence. That collection of jet drops, sometimes called the "marine droplet aerosol," is the main way in which particles of sea salt get into the atmosphere, for the jet drops can easily evaporate before falling back to the ocean's surface, thus giving up their salt content to the air. There's a less pleasant outcome, too. Bubbles that rise to the surface in stagnant lakes and ponds can likewise transfer harmful bacteria from the surface into the atmosphere, which the wind can blow anywhere.

CHAPTER TWO

Raining Cats and Dogs
Liquid Droplets in a Gas

Sing, as when we bob for shelter
out of the rain, and from our chorus
swell
hubble-bubble
to the
top.

—Aristophanes, "The Frogs"

A water drop in air is completely the opposite of an air bubble in water, as Figure P.1 shows. Rain is a seemingly never-ending cascade of water droplets, from the clouds above to the ground below. When we burn coal and oil, both of which contain trace amounts of sulfur dioxide, this gas can combine with water in the atmosphere to produce a mixture of water and sulfuric acid. The result, when the weather is bad, is what is called acid rain. But it's not just industrial plants that produce these corrosive droplets—volcanic eruptions yield the same result.

Another example of a liquid droplet in a gas is the ink-jet printer that PC users know and love: the printer sprays droplets of ink onto the surface of a sheet of paper. The challenge is for the droplets to retain their shape, and for the paper to "hold the ink," so that the liquid won't spread on the paper's surface, yielding a blotchy, botched, print.

We often hear the phrase "rock solid," but rocks aren't always solid. It seems that there are four major areas in the world, called strewfields, that contain tektites. These small rocks look like liquid droplets. Geologists think that meteorites smashed into the earth with such great force that they caused the rocks to melt, which sent liquid droplets of rock hurtling into the atmosphere. These droplets cooled as they headed back down to earth, creating the distinctive droplet shape characteristic of tektites.

London is known for fog, but perhaps we should call it London Smog. In 1952, a "pea-souper" descended on the city from December 5 to December 9. Fog is in fact a low level cloud, a collection of water droplets suspended in air. In those years, England's homes were heated by coal fires, producing a deadly yellow-black fog containing sulfur dioxide, coal particles, and many other nasty substances. Reports suggest that more than 4,000 people died, and tens of thousands more had health problems caused by the pollution. Luckily, the government realized that this was a major problem, and shortly afterward passed the Clean Air Act, which prevented future occurrences.

And for some home-grown fog, wait for a chilly fall day, go outside, and exhale. The air we breathe out contains quite a bit of water (and some saliva), and as the water vapor in our mouths hits the air, it condenses to form our own mini-fog. You can see your own breath!

Droplets can also be put to good service. Much swifter than the old-fashioned paint brush is the spray painter, which passes a stream of compressed air through liquid paint to produce a spray of tiny paint droplets. And in medicine, one of the best ways to deposit medication deep within the lungs, to help those with breathing problems, such as children with cystic fibrosis, is the same as spray painting: pass compressed air through the liquid antibiotics in what is called a nebulizer, and the minute droplets produced can be inhaled far down into the lungs, where they can do the most good. Avid gardeners use the same technique to keep their lawns green, using lawn sprinklers to send water droplets—artificial rain—across their backyard. (The technical term for liquid droplets suspended in a gas is, not surprisingly, an aerosol.)

Heavy rain can bounce off a wet road. But what's amazing is that the droplets don't ricochet off the asphalt itself, but off the water that already coats the surface. Water has an elastic skin (hold a glass of water up to eye level, and you can see the "skin" layer) and this, like rubber, has a certain tension. A raindrop hits the water's surface and compresses the elastic skin, which pushes back, launching the droplet back into the air. In *Why Does a Ball Bounce?* (Buffalo: Firefly Books, 2005), Adam Hart-Davis describes an experiment developed to show a bouncing droplet. Pour water into a big plastic box to a depth of about 1 inch (2.5 centimeters). Use an eye dropper, or something similar, to let a drop of water fall onto the water in the box. As it hits, the droplet deforms the surface. It makes a depression where it lands, but this is surrounded by a slight circular rim, or hill, made from the water pushed out of the depression. This rim cannot last, and as it collapses it forces water both outward and inward. The water moving outward forms the beautiful circular ripples you might see on the surface of a pond when you throw a rock into it. The water moving inward from, say, the left smacks into the water moving inward from the right, and this collision throws a drop about half an inch (1 centimeter) up into the air (Fig. 2.1). For an instant, the drop hangs poised, above a thin column of water, before falling back to the surface and sending out more ripples.

Not all droplets are equal. What's true for water is not true for milk, for example. If you let a droplet of milk fall onto a piece of black Perspex (Plexiglas or Lucite), it will form a little white pool. But a second drop

Figure 2.1. A drop of water striking the surface generates a vertical column of water and a second droplet suspended above the column. (Mustafa Deliormanli/iStockphoto)

Figure 2.2. The familiar "crown" created by a milk drop. (Eric Delmar/ iStockphoto)

falling into this pool doesn't splash the way water does. Instead, an enchanting crown of droplets cascades upward all the way round. The reason is that whereas pure water is a fairly simple substance, milk is an extremely complex fluid. It's a suspension of fatty acids in water, and is more viscous (stickier) than water, which prevents it from bouncing so freely. Not only that, but milk's different density and surface tension means that the bouncing, too, will be different. Last, milk is slightly inhomogeneous, so that its consistency varies from point to point; some parts are almost pure water, some are thick lumps of fatty solid—this is whole milk—and some are somewhere in between. The end result: a lovely photogenic crown (Fig. 2.2). For scientists and engineers, such beauty needs an explanation. What determines the height and breadth and form of the crown? Answers to these questions are still being explored, so perhaps there *is* some crying over spilled milk!

Faucet Physics

Scientists like to play, and what better place to do so than in the bathroom? Turn the faucet just enough to produce a skinny stream of water. If the faucet is high enough above the bowl, then before the stream touches down, it will break into droplets (Fig. 2.3). Next time you are in the bathroom, carry out the experiment for yourself. Breaking up is not hard to do: the surface of a liquid is like a membrane with a tension. This surface tension creates a surface energy, so that the larger the surface area of a blob of water, the more energy it possesses. You can imagine a small section of the stream of water: it's a bit like a cylinder. But now imagine replacing that cylinder of water by two spherical droplets. The two droplets have less surface area, and so less energy, than the cylinder does. And so, the cylindrical stream of water will start to break up into droplets. This reaction is known as the Plateau-Rayleigh instability. Belgian physicist Joseph Plateau noted that the stream breaks up into droplets once the length of the stream is about 3.1 times its diameter. John William Strutt, the third Baron Rayleigh, proved mathematically that the breakup happens when the stream's length is π (3.1415 . . .) times its diameter. Before doing so, the stream forms "capillary waves," until it breaks into drops.

Figure 2.3. A stream of water from a faucet, the stream eventually breaking up into droplets. (Dschwen/Wikimedia)

Figure 2.4. A thin stream of water bent by static electricity. (Photograph courtesy of Colleen Condon)

When a fine column of water falls from a jet, it is first a cylinder, then necks and bulges begin to form, and at last beads separate, and little drops can be seen. The beads also vibrate, becoming alternately long and wide, and the sparkling portion of a jet, though it appears continuous, is really made of beads that pass so rapidly before the eye that it is impossible to follow them.

Those who entered the bathroom to comb their hair can have yet more fun. Whip out a plastic comb and rub it vigorously on, say, a woolen sweater. This generates a considerable electrostatic charge. Hold the comb close to a thin column of water and voilà! The stream curves toward the comb (Fig. 2.4). The reason, incredibly enough, has to do with the shape of a water molecule. Each molecule of pure water, H_2O, consists of two hydrogen atoms that bond with just one oxygen atom. Roughly speaking, these form a V shape, with the hydrogen atoms at the tips of the V and the oxygen atom at the base. But oxygen is slightly negative relative to the hydrogen, giving the water molecule a negative area around the base and a positive area around the tips. So when you bring a positively charged comb nearby, the water molecule turns around so that its negative part points toward the positive comb (opposites attract) and the water heads over toward the comb. If the comb is negatively charged, the water molecule points its positive region toward the comb and, again, heads on over toward it. No matter what the charge on the comb, the water stream deflects. Try it!

A dripping faucet can be annoying, but this too is a rich source of science. Just as with the stream of water, if a large drop of water gathers at the faucet, it may start to form a neck, due to gravity. But this, too, depending on the faucet's size, might be unstable. The net result? The large droplet splits up, sending a big droplet downward, and a smaller droplet following like a puppy dog behind it, while the rest of the droplet snaps back upward to cling to the faucet. Figure 2.5 shows four successive stages in the process after the drop has become unstable. The smaller droplet produced is known as *Plateau's spherule*.

Even the droplets that *don't* fall are instructive. If you allow a liquid droplet to form in a highly controlled laboratory experiment, you can capture it in a high-resolution image. Let plenty of such drops fall into a beaker, measure the weight of the beaker, and you can infer the weight of one liquid droplet.

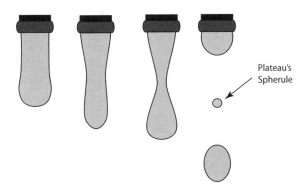

Figure 2.5. Four stages of a droplet that has become unstable. The smaller droplet is known as *Plateau's spherule*. (Greg Nicholl)

There's only one other item that affects the shape of the droplet, and that's the surface tension between the liquid and the air. So, armed with the theory that explains the shape of a liquid droplet that hangs under gravity, and knowledge of the droplet's mass, scientists can determine the approximate value of the surface tension. This is known as the *pendant-drop method.*

And as for the drip, drip, drip of the leaky faucet? It's unpredictable. What's more, it's a great example of the field of chaos research. Students can record the moment when a droplet is released from a leaky faucet and then measure how much time passes before a fellow droplet joins it. But don't expect a consistent pattern in these measurements. (A good reference is James Gleick, *Chaos: Making a New Science* [New York: Viking, 1987].)

Spider's Web

Who hasn't seen a spider's web? But take a closer look. Each strand of a web is a fine elastic thread covered with beads of a sticky liquid. In a good web there are over a quarter million of these beads, and it's the beads that catch the flies for the spider's dinner. It takes the average arachnid about one hour to make a whole web, and generally the spider makes a new one every day. (There is a fine example in the group of color photos.) Gluing each of the million beads into place would be exhausting and take far too long. The spider puts Plateau to enviable use. She spins a thread but at the same time coats it with a wet sticky liquid. This, at least to begin with, is also a thin cylinder. But just like the stream from the faucet, it won't be too long before the liquid breaks up into beads, without prompting.

Liquid-Drop Model of the Atomic Nucleus

Back in 1937, physicists were struggling to understand the properties of the atomic nucleus. One way to proceed is with an analogy, using a known piece of "off the shelf" physics to see how useful it can be. In a series of experiments, physicists determined that the density of an atomic nucleus is roughly constant; the liquid-drop model was born. Imagine that an atomic nucleus behaves roughly the same way the constant-density droplet of

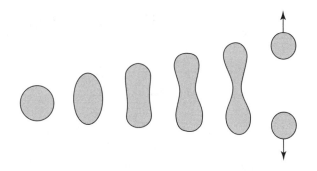

Figure 2.6. A nucleus splitting into two parts, by the process known as fission. (Greg Nicholl)

water behaves. It would have a surface tension that tries to keep the nucleus confined to a certain size. But under the right conditions, the nucleus can be split. A nucleus of uranium can be split into two parts (a process called fission), emitting energy, which can be put to use in a nuclear power station or a nuclear weapon. Figure 2.6 shows how this happens. The drop begins to vibrate, the oscillations becoming greater and greater. Eventually, as with the stream from the faucet, it breaks into two parts. This process emits vast amounts of energy, either for our benefit (as with nuclear power) or to our detriment (as in a nuclear weapon). A water droplet, on a hot day, will lose some of its water molecules as they evaporate into the warm dry air around it. The liquid-drop model for the atomic nucleus exhibits something similar: the equivalent of the water droplet's evaporation is the loss of part of the fluid of the nucleus to the environment outside, a process we call radioactive decay. The model worked reasonably well for a while, but physicists now have far more precise and mathematically far more powerful ways to describe the atomic nucleus.

The Swirl of Bath Water

One way to carry out faucet physics is to draw yourself a hot bath. When you've finished soaking luxuriantly, pull out the plug for one final test. Which way does water swirl as it exits the tub? Some people claim that it swirls clockwise in the Southern Hemisphere and counterclockwise in the Northern. This, though, is not true, as precisely engineered experiments (some were reported in *Scientific American*) and some appeal to higher-level physics will confirm. A more likely explanation lies in the physics of the Conservation of Angular Momentum. As you move your hand through the water to pull out the plug, even in a perfectly still bath of water, you will give the water a little swirl. It might be that a piece of soap is stuck in the drain—and the drain is usually near the end of the bath. No matter what the cause, the flow is asymmetric, meaning that more water will go into one side of the drain than into the other side. For example, look at a tub full of water swirling gently around the bath. As the water drains, less and less of it remains, but the angular momentum of the water's exit remains constant. Roughly speaking, because angular momentum is conserved, the less water there is, the rate at which it swirls must increase. In the end, it finally gurgles down the drain with very small mass and very large angular velocity. Whether it swirls clockwise or counterclockwise depends on the direction of the initial small movement of the water in the full bath, or the direction given to the water when the plug is pulled out. What better reason for taking a bath, than to carry out a series of swirling experiments!

Morning Dew

Few things are as pretty as dewdrops on rose petals. But how do they form? On a hot afternoon the sun warms both the ground and the atmosphere. The warmer the air is, the more water it can hold. So as the day progresses, lakes, ponds, streams, and swimming pools will all lose water through evaporation. This process increases the amount of water vapor in the air. But at night, the temperature may plummet. Just as wringing out a soaking wet facecloth causes water to squirt out, so the cooling of the air at night "squeezes out" the extra water that the atmosphere can no longer accommodate. If this condensation occurs high up in the atmosphere, the droplets form clouds, and if the clouds are sufficiently concentrated, they form rain. When condensation occurs near the ground, the droplets form an early morning mist or fog, which is often burned off when the sun warms it enough to turn the droplets back into vapor. Sometimes, during the night, the ground cools more quickly than the atmosphere. At such times,

Figure 2.7. Dew condensed on a leaf. (Wikimedia)

the water vapor condenses onto twigs, rocks, and leaves (Fig. 2.7). Dew, in fact, is usually purer than rainwater. As raindrops fall, they sweep up polluting gases, for example sulfur dioxide from power stations. Dew, however, is water that has condensed directly from the air, and has not been able to pick up pollution on the way, and is thus pure water.

If the ground temperature falls below the freezing point, ice is deposited, or dew already deposited freezes; the result is hoarfrost (Fig. 2.8). Some items in Mother Nature's cabinet don't conduct heat very well, and it's on these that hoarfrost will collect. Other likely candidates are objects with relatively large surface area, which can't store up heat energy readily. This means that leaves and blades of grass (though some of the observed moisture is produced from within and not deposited) will collect substantial deposits of hoarfrost.

Acid Rain

Air is a mixture of gases, four-fifths of it nitrogen. Under normal circumstances, nitrogen is a fairly inert element, loath to take part in chemical reactions. But under the high temperature and pressures inside the engines of cars and trucks, it can be persuaded to combine with the oxy-

Figure 2.8. Hoarfrost collected on branches of a maple tree. (Photograph courtesy of Greg Nicholl)

gen in the atmosphere to form nitrous and nitric oxides. Nitrous oxide, when used as an anesthetic, is known as laughing gas, but in the atmosphere it is no joke. These oxides dissolve in water, producing nitrous and nitric acid. Rain that falls subsequently is acidic, and this can be devastating to trees. A tree felled in 1989 in Germany's Black Forest revealed that the most recent rings were packed closely together, indicating that growth had been stunted in the preceding 20 years, owing to the enormous increase in motor vehicle traffic pollution in later decades.

Bubble Chambers and Atom Smashers

Ben Nevis, the highest mountain in Britain at 4,406 feet (1,344 meters), is the remains of a 400-million-year-old volcano, most of which has been eroded away by glaciers and Scottish rain. The walk up the mountain is quite long, because you start almost at sea level. On one side of the summit, a sheer cliff drops away to the north and has to be avoided—unless you are a mountaineer in search of a challenge!

Back in 1895, William Swann created his own challenge: he ran from the local post office up to the summit and back down again, taking about 2 hours 45 minutes to accomplish the feat. The race up Ben Nevis is now an annual event, the current record standing at about 1 hour 25 minutes.

The gentlest way to the summit is via the path constructed in 1883 by the Scottish Meteorological Society, which built an observatory on a plateau on top for studying the weather. Wind speed, temperature, and air pressure were recorded every hour for 21 years. There was also a small hotel on the plateau. When the weather was bad, supplies were hauled up by cable. The remains of the observatory, the hotel, and the cable lift can still be seen today.

In September 1894, just a few months before William Swann's sprint to the summit, one of the scientists who worked at the Ben Nevis Observatory, C. T. R. Wilson, became intrigued by the *glory* or *Brocken Spectre*. A glory is a faint rainbow around a person's shadow when the shadow is thrown by the sun onto a cloud below. (We'll hear more about this later.) In order to study this further, Wilson returned to the Cavendish Laboratory at Cambridge University and built a cloud chamber (a modern version of which can be seen in Fig. 2.9). In the chamber, air saturated with water vapor is made to expand suddenly. Just as with dew, this water vapor is ready to condense at the slightest excuse. So, if an electrically charged atom or group of atoms (ions) passes through the air/vapor mixture, the ions act as nuclei on which drops of water condense. The trail of water vapor marks the path taken by the ions. This proved to be a major step in experimental particle physics. The cloud chamber has been of great use in nuclear research and, for his invention, Wilson shared the Nobel Prize for Physics in 1927.

In 1951, Donald Glaser won the Nobel Prize for the invention of the bubble chamber. The story is told that he came up with the idea while watching the bubbles in a glass of beer in a bar in Ann Arbor, Michigan. The idea is similar to what occurs in a pressure cooker. If you put water in a pressure cooker and crank up the pressure, the water can boil at

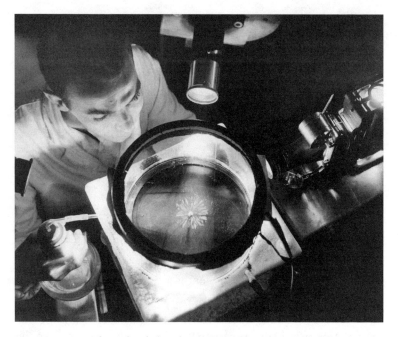

Figure 2.9. A modern cloud chamber. (National Aeronautics and Space Administration)

temperatures greater than 100°C. We describe this water as superheated. In a bubble chamber, a liquid is superheated, but the pressure is then suddenly released. If the liquid is extremely pure, then when the pressure rapidly drops, bubbles may not form in the liquid for perhaps 30 seconds or more. During this quiet period, if ionizing particles enter the liquid, they may cause bubbles to be formed, and the bubble trails, when photographed, show the paths of the ionizing particles. By this means, scientists can look at the trajectories of charged particles and hunt for ones that no one has previously discovered.

Tracks recorded by bubble chamber photographs are much clearer than their cloud chamber counterparts, and they can be taken more rapidly. Figure 2.10 shows the Big European Bubble Chamber (BEBC) at the European Centre for Nuclear Research, best known by its French acronym CERN. The chamber can be filled with liquid hydrogen or a mix-

Figure 2.10. The Big European Bubble Chamber (BEBC) at the European Centre for Nuclear Research. (Photograph © 1971 CERN)

ture of neon and hydrogen. The chamber is surrounded by superconducting niobium-titanium coils (not visible in the photograph), which produce an intense magnetic field inside the chamber. The magnetic field is a key ingredient, for a charged particle in a magnetic field will describe a circle. What's more, the magnetic field will bend a positive-charged particle in, say, a counterclockwise circle, but bend a negatively charged particle in the opposite, clockwise direction. And the radius of the circle that the particle describes depends on how heavy the particle is. Uncharged particles move in straight lines. So, when presented with intriguing photographs of particle tracks, physicists can determine the charge and mass of the particles that have been created. It's by means such as this that the giant atom-smashing devices discover brand new elementary particles. In Figure 2.11 we see a photograph of a neutrino

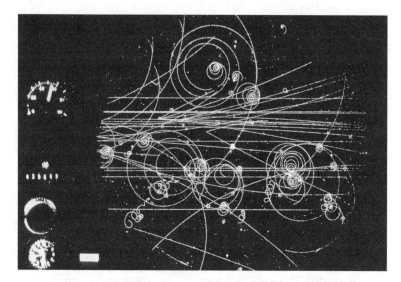

Figure 2.11. A neutrino reaction recorded by the Big European Bubble Chamber. (Photograph © 1970 CERN)

reaction taken in the Big European Bubble Chamber of CERN at Geneva. (We needn't attempt to explain just what is going on in there!)

CERN is currently conducting the biggest experiment in the history of science. A team of 10,000 scientists from 80 countries has constructed the Large Hadron Collider (LHC). Two beams of protons are sent whizzing around the 17-mile-long collider in opposite directions, at speeds that are 99.9999991 percent of the speed of light. The collider tunnel is about 300 feet underground and is about as wide as a New York Subway or London Underground tunnel. But this is where scientists hope to detect the Higgs boson. A by-product of a reaction that may take place when protons collide, the Higgs boson is a particle that (if it truly exists) explains why matter has mass. A Nobel Prize in physics hinges on the outcome.

Clouds

Clouds, which consist of water droplets or ice crystals, are classified into four main groups, or families, according to the heights at which they form (Fig. 2.12a–d). But each family is further subdivided into classes, representing ten fundamental cloud types. *Cirrus* (or the prefix *cirro-*) indicates high clouds; *alto,* medium height; *cumulus* (*cumulo-*), accumulated isolated patches or heaped formation; *stratus* (*strato-*), a layer of cloud; and *nimbus* (*nimbo-*), a heavy type of rain cloud. Air is not perfectly dry—in nature, it invariably contains some amount of water, and this is what causes humidity. The amount of water that the air can contain depends on the temperature. So, if the air temperature falls below a certain value (the *dew point*) the extra water vapor in the air condenses out as small droplets of water (or, at low temperatures, small ice crystals)—forming a cloud. Condensation takes place whenever saturation is reached, for there are always impurities in the air (such as dust or soot particles, or salt from the sea) that can serve as the nuclei around which droplets can form.

Water droplets in a cloud fall slowly to the ground. The braking effect, determined by air resistance, depends on the droplet speed: double the droplet's speed and you double the air resistance. A truce is reached, when the droplet's weight is exactly counterbalanced by the force of air resistance. From that point on, the droplet falls at a constant speed, known as its *terminal velocity.* This is true for small drops. But water droplets greater than 5.5 millimeters in diameter are broken up into smaller drops as they fall, so 5.5 millimeters represents the greatest possible diameter for a raindrop, and the highest terminal velocity is about 8.0 meters per second. This is the raindrop's speed *relative to the air;* so if there's a vertical upward current of air rising at 8.0 meters per second, raindrops of this diameter remain poised at a constant height above the ground; an upward current of air exceeding this value will carry them upward. The chief difference between a cloud and a fall of rain lies in the rate of settling relative to the ground.

A cloud whose droplets are 0.01 millimeter in radius will settle slowly, at first. But these droplets may collide and coalesce, falling faster as they grow, and collecting more small drops as they fall. By the time they have reached a diameter of a millimeter or two they may fall as rain, with the considerable terminal velocity of 4 or 5 meters per second, but if there is an upward current of sufficient velocity, they will stay up as a cloud that continues to grow.

Figure 2.12 (above and opposite). Four types of clouds: (a) cirrocumulus, (b) altostratus, (c) stratocumulus, and (d) cumulonimbus. (National Oceanic and Atmospheric Administration)

Thunder and Lightning

Few things in nature are more spectacular than a storm that produces thunder and lightning. These impressive events are caused by friction. Clouds are rising or falling and, as they pass neighboring clouds, may literally rub against them. The friction between air and raindrop can cause the droplet to become electrically charged (Fig. 2.13). Air is a fairly good insulator of electricity (which is why high-tension power lines are relatively safe), but if the cloud becomes highly charged, then the voltage difference between the cloud and the earth can be extremely great. Once the difference exceeds three million volts per meter, the cloud will discharge an electric spark to the ground, which is what we call lightning (Fig. 2.14). To make buildings safe, tall metal poles ("lightning rods") are attached to them. The poles attract lightning, sending a voltage down the pole and into the ground and not through the building, where people may live. (Credit for inventing the lightning rod is usually given to Ben Franklin.)

Lightning, though, is often accompanied by thunder. Lightning—whose color is determined by the gases the air is made of—can be seen straightaway, for light travels at 300 million meters per second. The bang of thunder, though, trundles along at a far more sedate pace, only 300 meters per

Figure 2.13. Friction between air and raindrops can cause rain to become electrically charged. (Greg Nicholl)

Figure 2.14. Lightning touching the ground. (National Oceanic and Atmospheric Administration)

second. So, what you do is start counting seconds as soon as you see the lightning, and stop when you hear the thunder. If 5 seconds have passed, then the lightning took place 5 × 300 meters away, which is 1,500 meters or, for the nonmetric, about a mile. That gives us an easy way to guess the distance: count the number of seconds between *flash* and *bang* and divide by five: this is the distance of the storm you're watching in miles.

Snow

Mountains are often covered with snow, and if you've ever been skiing or mountaineering you know it can also get pretty cold up there. The reason for those frigid conditions is that the temperature of the atmosphere drops about 6°C for every 1,000 meters rise. So if the lowland temperature in Alaska happens to be a balmy 15°C (59°F), then the top of Mount McKinley (the highest peak in the United States at 6,194 meters or 20,320 feet), will be about 6 × 6.194 = 37°C cooler. Climbers who reach the summit are enshrouded by air whose temperature is −22°C (−7.6°F). On that same 15°C day in Anchorage, the air temperature reaches the freezing point 2,500 meters above the ground. This means that all except the lowest clouds are, on the whole, below freezing. They don't *have* to consist of ice crystals, because small water droplets can be supercooled far below the normal freezing point yet remain liquid; but larger drops freeze more readily. Low clouds, up to a height of about 3,000 meters, usually consist entirely of water droplets, whatever the temperature, provided that the droplets are small. But if the droplets are cold enough and also large enough to fall, freezing occurs and we get snow. The beautiful and varied forms of snowflakes are formed by the joining together of the small crystals of ice formed from individual droplets (Fig. 2.15). Snowflakes are so pretty that they inspired Johannes Kepler, best known for his three laws of planetary motion, to write a book *De Nive Hexangula* (*The Six-Cornered Snowflake*) about them. The book, published in 1611, is celebrated among mathematicians, for it contains an assertion, the Kepler Conjecture, that they have spent 400 years trying to prove or disprove.

Figure 2.15. The ice crystals of a single snowflake. (National Oceanic and Atmospheric Administration)

Clouds up to about 4,500 meters above sea level contain mainly supercooled water, though some ice crystals are present, owing to the freezing of the larger drops. The high cirrus clouds above 6,000 meters consist mainly of ice crystals. The amazing structure of these clouds is best seen from an airplane as it flies into, or just above, them.

Hail

Hail usually occurs only in the most violent disturbances of our atmosphere, and is usually associated with thunderstorms. The formation of hail is not a simple process. The first step is to have a large droplet, big enough and high enough to freeze. For the process to continue, it mustn't fall, so there needs to be a strong updraft. The baby hailstone needs to grow, which it can do in two ways: by collecting small supercooled cloud droplets, which freeze on it instantaneously and form the loose white, roe-like structure of soft hail; or it can sweep up droplets that are just below freezing, which spread over the hailstone's surface and freeze slowly, forming a solid shell of clear ice. In the violent air currents of a thunderstorm, the hailstone may be swept upward by ascending air currents and then plummet down to the lower part of a cloud several times before it becomes large enough to fall to earth. A section cut across such a hailstone will show successive rings of white ice and clear ice, much like the layers of an onion: the white layers are caused by supercooled water and ice crystals that join the hailstone; the clear ice is due to the water droplets that were above the freezing point before they hit the hailstone.

Hailstones, unlike raindrops, do not break up while falling, so they can be of wide girth. The record books report hailstones the size of grapefruit, about 5 inches (13 centimeters) in diameter and weighing up to 2 pounds (1 kilogram). A committee of experts convened to discuss the largest hailstone ever to land in the United States, which touched down at Aurora, Nebraska, in 2003—it was over 7 inches in diameter. If that's not enough to capture your attention, the terminal speed for an object as dense as a hailstone is not the paltry 25 feet per second of a raindrop; hailstones can crash to earth at over 100 miles per hour, with devastating effect.

Figure 2.16. A vapor trail created by an airplane at high altitude. (F. Ronald Young)

Vapor Trails

High-flying aircraft sometimes create vapor trails, often persisting for many miles. These are caused by the great heat of the engines and the consequent condensation of water vapor in the cooled air behind the aircraft (Fig. 2.16). But aircraft are not the only transient vehicles to do this: curiously enough, infrared photographs of the oceans taken by the American space shuttles show similar long lines. These lines, which can be hundreds of miles long, mark the wakes of ships. As the ships plow across the oceans, their propellers churn up plankton. These minute lifeforms give off molecules known as surfactants (short for surface-active agents), each of which has a water-loving side and a water-hating side. These molecules are carried up to the surface by the cavitation bubbles produced by the ships' propellers. This biological material, which forms a film on top of the water, appears as the long lines when photographed in infrared light.

The Rainbow

The rainbow comes and goes,
And lovely is the rose . . .
—*William Wordsworth,*
"Ode: Intimations
of Immortality"

Everyone's seen a rainbow. But if you're with a crowd of people on a rainy, yet sunny day, one of them may shout "a rainbow" and yet you see nothing. You turn around and there it is, the enthralling arc in the heavens. This happens because, to see a rainbow, you need to have your back to the sun. (On an exam paper I once marked, the student wrote that the rainbow was on the observer's backside!) The poetic beauty of the rainbow is caused when light from the sun undergoes total internal reflection in the uncountable numbers of raindrops and is then refracted into the colors of the spectrum (Fig. 2.17a). White light strikes the front of the droplets and is bent (the extent of the bending depends on the wavelength of the light). The rays of colored light then strike the backs of the droplets and head back toward the observer, just as light from anything else we see does. But as the light rays strike the front of the droplet on their way out, they are deflected once again. And as the diagram indicates, the red rays are deflected downward more than the violet rays, yielding the familiar spectrum of colors.

Sometimes, you can see two rainbows in one. In a *double rainbow*, a secondary rainbow is formed above the primary one by rays that undergo *two* internal reflections (Fig. 2.17b). In the secondary rainbow, the violet rays are deflected downward more than the red rays, with different results.

Think now of a droplet high up in the (single) rainbow. As the red rays are deflected downward more sharply, you will see red rays only from the upper droplets; the violet rays will miss your eye. Droplets at the bottom of the rainbow deflect their red rays downward and these rays won't enter your eye; only the uppermost rays, the violet ones, will do so. So, in a single rainbow, the topmost part of the arc will be red, the

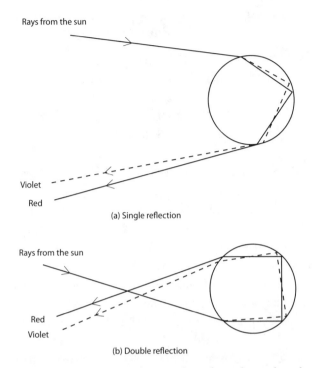

Rays from the sun

Violet

Red

(a) Single reflection

Rays from the sun

Red

Violet

(b) Double reflection

Figure 2.17. Light enters into raindrops and is refracted into the colors of the spectrum to create (a) a single rainbow (by a single reflection) or (b) a double rainbow (by a double reflection). (Greg Nicholl)

bottom violet (Fig. 2.18). The order of colors of the rainbow (or any spectrum from sunlight) is remembered in Britain by the mnemonic

Richard of York Gave Battle in Vain
Red Orange Yellow Green Blue Indigo Violet

or, in America, by the acronym "Roy G. Biv," which has been used as the title of a couple of songs. Don't forget, though: in a double rainbow, the colors of the secondary, weaker rainbow are reversed.

Figure 2.19 shows a rainbow in the late afternoon. This, and early morning, are the best times to see a rainbow, because the sun must not be too high in the sky. To sharpen your observing skills, see that the sky inside the rainbow is bright, because raindrops there also reflect light

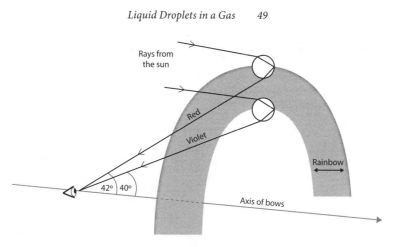

Figure 2.18. Rays reflected downward so that the topmost part of the rainbow is red and the bottom violet. (Greg Nicholl)

Figure 2.19. A rainbow in the late afternoon. (Photos.com)

directly back toward you. Rainbows are rarer than might be thought. In any one place in rainy England there are fewer than ten bright ones in a year.

For a real challenge, enough light can be reflected by the moon to generate a lunar rainbow, also low in the sky, but these are extremely rare. Something far more common is a rainbow produced from drops of sea spray, or one formed from the spray of a garden sprinkler, at the right time of day.

Halos

Rainbows are what we get when light passes through water droplets in the atmosphere. But higher up, there are no droplets, only ice crystals. These too, though, can create stunning visual effects: halos. Tiny ice crystals are flat-based hexagonal prisms, and light is refracted as it passes through them (Fig. 2.20). The ray enters one face and leaves through another, angled at 60 degrees to the first. If all the ice crystals pointed in the same direction, there would not be much of interest to observe. But because the crystals are arranged at random, a circular halo forms around the sun and this can be photographed. (You should never look directly at the sun.) If the center of the sun is dead ahead, at zero degrees with respect to the camera's view point, then a ring formed by the ice crystals will be at an angle of 22 degrees. (Figure 2.21 shows a complete 22-degree halo.) The halo, like the rainbow, has colors, but they are weak. In certain calm conditions, the ice crystals

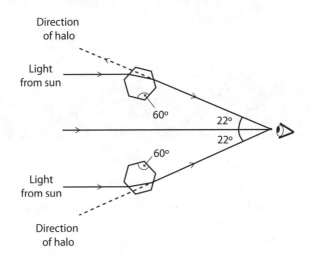

Figure 2.20. Light refracted through flat, hexagonal ice-crystal prisms to create a halo. (Greg Nicholl)

A soap bubble suspended in midair. (Mila Zinkova/Wikimedia)

Top, A drop of water striking the surface generates a vertical column of water, a second droplet briefly suspended above it. (Mustafa Deliormanli/ iStockphoto). *Bottom*, The "crown" created by a milk drop striking the surface. (Eric Delmar/iStockphoto)

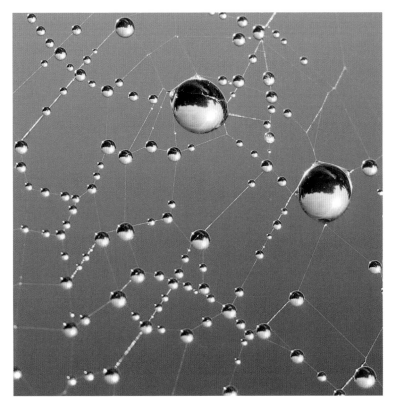

Droplets of morning dew suspended on the strands of a spider web.
(Luc Viatour)

A double rainbow appearing over the Virgin Gorge in Utah. (Photos.com)

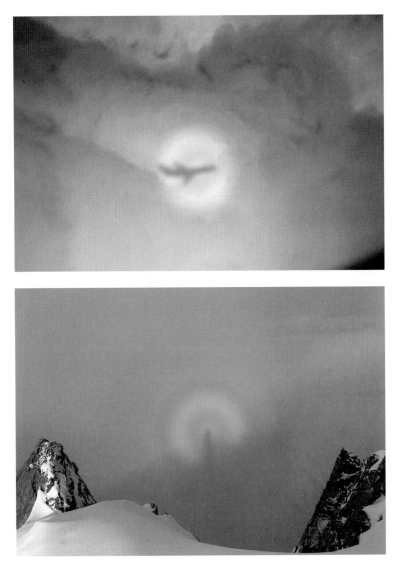

Top, A solar glory as seen from an airplane. (Mila Zinkova/Wikimedia). *Bottom*, A Brocken Spectre created when the shadow inside the rings of a glory is distorted. (Dmitry N. Zhukov/Wikimedia)

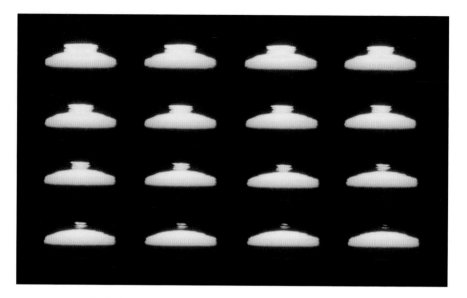

A green flash from a setting sun. (Mila Zinkova/Wikimedia)

Honeycomb is perhaps one of the tastier examples of a solid foam. (Photos.com)

The anti-bubble, a sphere of soap solution, surrounded by a thin shell of air, surrounded by the bath of soap solution. (Photograph courtesy of Terry Fritz/antibubble.com)

Figure 2.21. A complete 22-degree halo. (Wikimedia)

may not be arranged at random, but may all float with their axes vertical, oscillating gently about this stable position. The only visible portion of the halo then will be that part of it that is at the same height above the horizon as the sun and only streaky blobs of light in directions at 22 degrees with the sun will be visible. These are known as *parahelia, sun dogs,* or *mock suns.* They are more common in the cold climates of the Arctic and Antarctic.

The halos produced by ice crystals should not be confused with the much more common "ring round the moon." This luminous ring, called a *corona*, is caused by the scattering of light from the sun or moon by small water droplets of uniform size in the earth's atmosphere. The size of the droplets determines the angular diameter of the ring, whereas for halos the angular diameter is always the same.

Glories

Whenever a mist or a cloud is below you and the sun breaks through to shine, look for a glory. A good place to spot them is on mountains and hillsides, from aircraft, or in sea fog. Glories, which are formed when light is scattered backward from individual water droplets, have a bright center surrounded by even brighter rings of light. The rings are delicately colored, blue on the inside changing through greens to red and purple outside. Sometimes, three or even four sets of rings are visible. Glories are nearly always accompanied by the observer's shadow, or that of the aircraft you are sitting in (Fig. 2.22). Glories are often seen from aircraft, but you must be above the clouds. On your next plane flight, get a seat opposite the sun and away from the wing and watch for glories around the aircraft's shadow.

The shadow of the observer inside the rings of a glory, often grotesquely distorted by perspective, is called a *Brocken Spectre* (Fig. 2.23). Its name comes from its frequent appearances on the Brocken, the highest peak in the Harz Mountains of central Germany, where the display was called the spectre (ghost) of the Brocken. This spectre, or *fog-bow*, was observed by Edward Whymper when descending the Matterhorn in 1865 after the first successful ascent (Fig. 2.24). One of his party had slipped and dragged the

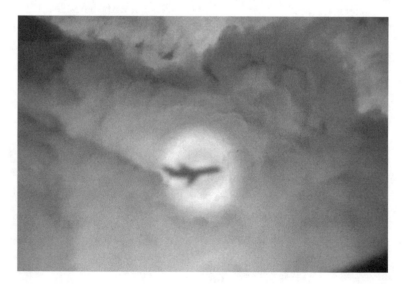

Figure 2.22. Solar glory as seen from an airplane. (Mila Zinkova/ Wikimedia)

Figure 2.23. A Brocken Spectre created when the shadow inside the rings of a solar glory is distorted. (Dmitry N. Zhukov/Wikimedia)

others down. The rope broke and four members of the party fell to their death. Whymper and his two guides, the Taugwalders, survived and saw the fog-bow. In his book *Scrambles amongst the Alps*, published by John Murray (famously the publishers of Lord Byron's poetry) in 1871, Whymper reports:

> The sun was directly at our backs; that is to say, the fog-bow was opposite to the sun. The time was 6.30pm. The forms were at once tender and sharp; neutral in tone; were developed gradually, and disappeared suddenly. The mists were light (that is, not dense), and were dissipated in the course of the evening.
>
> It has been suggested that the crosses are incorrectly figured in the illustration, and that they were probably formed by the intersection of other circles, or ellipses as shown in the annexed diagram. I think this suggestion is very likely correct; but I have preferred to follow my original memorandum.

Figure 2.24. Woodcut of a Brocken Spectre, based on the fog-bow observed by Edward Whymper when descending the Matterhorn in 1865.

The Green Flash

Imagine a luxurious vacation on a tropical island. The sun slips slowly downward to a clear ocean horizon. Ever reddening as it sinks, soon only a thin sliver of the once-hot disc remains. Then, as that too shrinks and diminishes, it shines forth for just a moment in a most vivid emerald green. Then the green is gone, and only the dark ocean remains. This is the green flash shown on page six of the color gallery. The green flash, strangely enough, is evidence that our planet is spherical. The earth is cloaked by an atmosphere, and the atmosphere can bend light. The amount that the light bends depends on its wavelength. The sun sets and goes below the horizon, and as it sinks, the red rays are increasingly bent by the atmosphere. These are the rays our eyes detect. As the sun sinks farther, the atmosphere can no longer bend the red rays to meet our eyes; but it *can* curve the green light from the sun to meet our eyeballs in just the instant before no direct sunlight can reach us at all.

But if the atmosphere bends the orange of the setting sun and then the green, why don't we then see a blue flash? The answer is that our atmosphere scatters blue light most of all (which is why the sky is blue), and light from a blue flash almost never makes it onto our retinas or camera lenses because it's so thoroughly scattered along the way. But, in a dry, pollution-free sky, on rare occasions, you can sometimes see a blue, or even violet, flash.

If you are energetic and are observing the green flash while standing on a hill, if you then run up higher after the flash has occurred, you might see another flash! There are sunrise flashes, too, also green, if you are alert for the very place and moment that the sun will appear.

The green ray figures in Osbert Sitwell's travel book *Escape with Me* (London: Macmillan, 1939). While traveling aboard a French ship from Marseille to Saigon in 1937, he writes:

The afternoons, then, become increasingly restful, though tea-time always brought a small influx to the bar and, just before sunset each evening, trouble would begin about Le Rayon Vert, the Green Ray. Recently there had been correspondence in the columns of *The Times*, and also, presumably, of certain French journals, concerning this mysterious and, indeed, mythical ray. . . . Accordingly, at the correct hypothetical hour, all the elderly gentlemen on board with any claim to scientific knowledge—and the French elderly gentleman who does not make such a claim for himself has surely never been born—would arrive on deck in a very important manner, carrying, hung round them, telescopes, opera glasses, cameras, compasses, theodolites and any other instrument they could find which had the right look about it, and dragging after them tripods, camp-stools, and metal stands: then, bridging their noses with special pairs of dark spectacles, they would proceed to stare back fixedly, but with a certain air of questioning and defiance, straight into the eye of the setting sun. After that, of course, green rays were visible to them in every direction, but, since each of these scientific enquirers had observed his particular one in a different quarter of the sky, this introduced into the data an element of doubt which led to an immense amount of quarreling and bickering, fomented by the ladies, who, of a sudden, openly took sides in this matter.

The Blue Sky

Green flashes, sun dogs, and double rainbows provoke amazement. In everyday life, though, we usually see an azure sky—but why? The sun emits all the colors of the rainbow. When sunlight reaches our atmosphere, it interacts with the molecules in the air. The molecules absorb some of the light from the sun, and scatter it in different directions. This process, known as Rayleigh scattering, was named after John William Strutt, third Baron Rayleigh, who first explained it. Not all light is equal: because our atmosphere is made up mostly of nitrogen and oxygen, the shorter wavelengths of light are the ones most scattered. Because the shortest wavelength of visible light is blue, our sky appears blue.

When the sun is overhead, its rays must penetrate only about ten miles of atmosphere to reach us. But when we are observing sunrise and sunset, when the sun is close to the horizon, its rays take a far longer journey through the atmosphere. This means that there is far more opportunity for the blue light emitted by the sun to be scattered by molecules in the air, preventing it from reaching our eyes. Because the sun gives off white light, the subtraction of its blue component leaves us a setting or rising sun that looks red. Other contributors to a red sky at night, in addition to air molecules, are particles of soot or dust. Forest fires often produce vivid red sunsets, as do areas of high air pollution.

The moon, though, has no atmosphere, so there is no such scattering and the sky looks black, as photographs of Buzz Aldrin walking on the moon will testify. Distant hills on the earth during daylight appear blue, and not green, for the same reason. The light reaching the observer from the hills is light that has been scattered at a wide angle. It contains, therefore, proportionally more blue, and that is what we see.

CHAPTER THREE

The Perfect Pint

The Science of Foams

The trick is to sell the customer a product which is mostly air.

—*Confectionery manufacturer*

The utilization of foam precedes the human race.* The nymph of the froghopper, after hatching from the egg, secretes a froth whose volume is many times that of the nymph, or that of the insect it will become, and the nymph then lives under this froth for several weeks, drawing sustenance from the plant beneath the foam.

Scientifically speaking, foams are collections of bubbles, each bubble consisting of a film that surrounds a tiny air space. Figure 3.1 shows bubbles that have come together on a liquid surface. There are many ways to produce a simple foam. One way is to blow a gas through a thin nozzle into a liquid. The classic example of this method is a child blowing air through a straw into a glass of chocolate milk. There's lots of noise, which is very appealing, and some satisfyingly large bubbles on the milk's surface. Or, instead, you could use the kitchen method of shaking or beating a liquid to fold gas into it. A traditional English concoction is the syllabub, which dates back to the time when the Tudors ruled the country. It calls for cream, sugar, wine, and lemon juice to be beaten together to form a delightful frothy dessert.

---

* This chapter owes much to the excellent book *The Physics of Foams*, by Denis Weaire and Stefan Hutzler (New York: Oxford University Press, 1999), to which the reader is warmly referred.

air

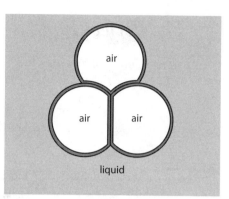

liquid

Figure 3.1. Bubbles collected on a liquid surface to begin the creation of a foam. (Greg Nicholl)

Another dessert famous for foam is ice cream. Before she became a politician, former British Prime Minister Margaret Thatcher worked for an ice cream manufacturer. The team she worked with came up with a method to introduce more air into the product. This gave the ice cream a lighter texture, popular with consumers, but also resulted in less ice cream (which is relatively expensive) and more air (which is relatively cheap), creating bigger profits. Mrs. Thatcher's method of course spread widely.

Beer shows another way, when gas bubbles nucleate in a liquid. In the head on a freshly poured glass of beer, the individual bubbles are of various sizes.

A final way, beloved of engineers, is sparging, which consists of blowing gas (usually an inert gas like helium or argon) into a liquid. This may sound unromantic, but sparging is a great way to remove unwanted, possibly dangerous, chemicals from a liquid that you need to use. Pumping the inert gas through the liquid forces the liquid to yield almost all of the gases that were dissolved in it, and these gases then froth up to the surface. The result is a nice clean liquid with a toxic foam, one that can be skimmed off and treated safely. Sparging is one of the many pro-

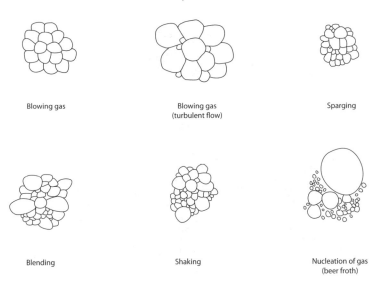

Blowing gas   Blowing gas   Sparging
(turbulent flow)

Blending   Shaking   Nucleation of gas
(beer froth)

Figure 3.2. Various ways of making foam. (Greg Nicholl)

cesses that can be used to make our waste water ready for drinking again. Figure 3.2 illustrates the various ways of making foam.

Foams are not in equilibrium. Three processes are always present, and the three always fight for control. First there is *drainage*—the liquid simply drains out of the foam (Fig. 3.3). Then there is *coarsening*, in which the bubbles grow bigger as time passes. Last, there is *collapse*, brought on when the liquid films between adjacent exposed bubbles break.

Liquid Foams

Most liquid foams do not last long. Drainage, which sees large bubbles rise while small bubbles remain beneath, reduces the film's surface while evaporation reduces it further. Surfactants, too, play a role. These molecules are nature's extroverts, each sporting a water-loving end and an oil-loving end. The grease on your dishes will not wash off with water, but a surfactant (detergents are full of them) will attach one end to the water, one end to the grease, and voilà, your plate is clean. If surfactants

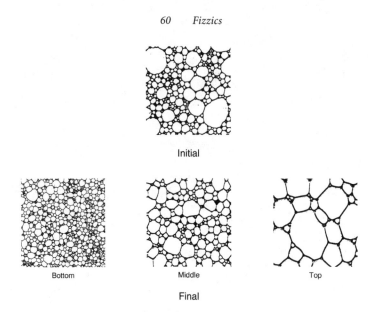

Initial

Bottom   Middle   Top

Final

Figure 3.3. Drainage of foam in a vertical column. As time passes, the foam is coarser at the top of the column than at the bottom. (Adapted from an image from Denis Weaire and Stefan Hutzler)

are not sufficiently concentrated in the foam, it will be short-lived, and impurities and additives (such as anti-foaming agents) will decrease its life span.

Liquid foam is most familiar in cleaning agents, soaps, shaving products, or beverages. The advantage of shaving cream is that the clearly visible foam clings to the skin of the shaver, guiding the way to a thorough shave, and the foam layer helps the blade glide smoothly over the surface, reducing the risk of nicks and scratches. Foams are popular in industrial settings, too. They carry chemicals efficiently and reduce the quantities of water needed. The textile industry, for example, employs processes involving cleaning, dyeing, or printing. Each requires chemicals to act over a large surface area of fabric, and foam carriers help reduce the amount of waste involved. Likewise, foams are used to decontaminate material inside a nuclear reactor, which speeds up the process when the reactor is to be shut down.

At the top of a glass of beer, bubbles collect in a foam, the head, which many beer drinkers will tell you never lasts long enough, though others argue that it lasts too long! For brewers, the appearance and staying power of the head are extremely important. Some beer manufacturers fatten the foam by putting additives in the beer. The idea is that air above the beer in a sealed can will deteriorate the beer. If the air is displaced by foam, this effect is reduced. A decent foam in beer can be achieved by the addition of natural proteins, which are natural surfactants or—less appetizing—by detergents. Sometimes, though, extra surfactants must be added. For some beers, nitrogen (in addition to carbon dioxide) helps create a fine, stable foam.

The foam in cans of Foster's beer is created by using a *widget*. A widget is a small plastic device measuring about 3 centimeters in diameter and containing a single small hole. The widget is placed into the empty can at the first stage of the packaging process. The can is then filled with beer, pressurized with nitrogen, and sealed. When opened, the can pressure immediately drops to atmospheric pressure. This release in pressure makes the widget spin, which in turn jets the gas down through the beer in the can, creating the draught foaming effect and the creamy head. An added benefit of the widget is that opening the can even at room temperature still won't make the beer gush, should the drinkers prefer their beer unchilled. (Readers should note that beer is best served at 6°C to 8°C.)

Soaps (solid or liquid), shower gels, or other bath foams are of petrochemical origin. A soap that foams easily on the hands or body presents a better commercial image (it sells more easily) than one that does not, although cleaning efficiency does not necessarily coincide with appearance. An abundance of foam may be welcome in products destined for body care, but in washing machines it is strongly advised against. My clothes had always been washed by a machine where, through the porthole, you could see nothing but soap suds. All went well until one fateful day when the outside drain was finally blocked by all the solidified soap residue from previous washes. The room flooded!

Candido Jacuzzi, who died in 1986, invented jet and submersible pumps, from which evolved the Jacuzzi whirlpool baths. The jets in a Jacuzzi produce a mixture of air bubbles and water. The result is a strong hydro massage that increases body circulation, and is thus beneficial for sufferers from rheumatism and arthritis. The jets can produce a silky-soft foam, or an aerated bubbling stream, or a pulsating shower. Most

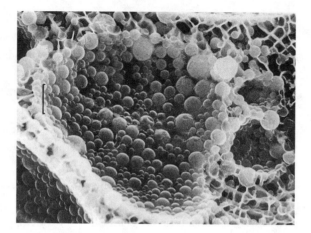

Figure 3.4. Emulsion droplets at the surface of an air bubble in a whipped cream. (Unilever, Colworth House, North Bedfordshire)

bathrooms do not have a sufficiently high head of water to produce the necessary power jets, and a special shower pump in the Jacuzzi increases the pressure of both hot and cold waters to the shower head.

The process of introducing air bubbles into a liquid by whipping or beating, which produces a foam, is complicated. Egg whites can be whipped into a foam, which lowers the density of such otherwise ponderous preparations as soufflés and sponge cakes. Figure 3.4 is an electron microscope photograph of emulsion droplets at the surface of an air bubble in a whipped cream. The air is the large shape occupying the main part of the picture. The emulsion droplets are the little balls. The bar at left represents five thousandths of a millimeter. A more complex and more variable structure is that found in ice cream, which is both an emulsion (droplets of one liquid dispersed in another) and a foam, and contains ice crystals as well.

Fire-Fighting Foams

Foams have been used to fight fires since the early 1900s. Generally, a fire needs fuel, oxygen, and heat in order to spread. Once this fire triumvirate is broken by removing one of these three constituents, the fire will cease. Fire-fighting foams may attack all three factors, by excluding oxygen from the combustion zone, by cooling the fuel below its ignition point, and by trapping the fuel vapor at the liquid surface. These foams are particularly useful for extinguishing aircraft fires.

In 2006 at Boundsfield, near Hemel Hempstead in the United Kingdom, at least twenty large tanks were set ablaze from a gasoline leak. It took the Hertfordshire Fire Brigade three days to accumulate enough foam from neighboring brigades to fight the fire.

Because the foam in a fire extinguisher is of low density, it forms a blanket that floats on top of the burning liquid. Water, by contrast, would merely agitate the liquid and help to spread the fire. If the fire starts in containers, the foam can be pumped to the bottom, from where it rises to the top and extinguishes the flames. But even when the fire is extinguished, the foam blanket should remain intact as long as possible, in order to reduce the risk of a fire breaking out again. This means that a slowly draining foam is desirable. The foam needs to be heat-resistant, which favors foams with a high liquid, as compared to gas, content. But the more liquid a foam contains, the less area the foam can cover in a given time. So the actual fire situation dictates which foam should be used.

Fire-fighting foams are classified according to their *expansion ratio* into low (from 5 to 1 to 20 to 1), medium (up to 200 to 1), and high expansion foams (up to 1,000 to 1). Low- and medium-expansion foams are produced by the use of a branch pipe, a device that aerates a foam solution. In other words, these are high-liquid, low-air foams. These can be sprayed up to 20 meters (more than 60 feet). Increasing the expansion ratio reduces the heat resistance, but allows a greater area to be covered quickly. And because the foam is lighter, it settles more easily onto a fuel surface without disturbing it. High-expansion foams (low-liquid, high-air) are produced by spraying the foam solution onto a net or gauze through which air is drawn or blown. But because of their light weight, high-expansion foams cannot be projected any reasonable distance, and must be applied directly to the fire.

Most buildings employ foam fire extinguishers, because the foam can quickly fill large spaces and won't cause problems if an electrical fire is involved (for which adding water would be a really bad idea). The dryness of the foam enables people who have become covered by it to continue to breathe normally. Conversely, the light weight of such foam can cause problems for outside applications, where it might simply be blown away by the wind. It also drains quickly and has little heat resistance.

Fire researchers continue to seek constituents for the perfect fire-fighting foam. Over the years, many different chemicals have been tested, including sodium bicarbonate (a vital component of any chef's kitchen) and aluminum sulfate dissolved in water. Foams are tested for their fire-fighting effectiveness by setting up a small fire with defined spatial extension. Researchers spray foam on the fire and measure how long it takes for the foam to reduce the area of the fire by nine tenths. The foam that achieves this in the shortest time is the better one. Other tests include determining the *burn-back time*, which measures how long it takes for the fire to re-ignite itself.

Honeycombs and Mattresses

Solid foams are just as important as liquid foams. If you've ever eaten honey, then you've benefited from a solid foam: honeycomb, the mass of hexagonal wax cells built by honeybees in their hive to contain their brood and stores of honey. Instead of surrounding a gas by a liquid, solid foams encase a gas in a solid. Another case in point, popular with children, is bubble wrap, which makes an addictive "pop" when the solid plastic film encasing the bubbles is ruptured. And for those soaking in the tub, sea sponges are also examples of solid foams.

Bread, too, is a common solid foam. In bread making, the addition of yeast produces bubbles of carbon dioxide in the dough, which give it a light airy texture, otherwise known as leavening. Mere starch alone is not elastic enough to hold the bubbles, which can burst. So it's just as well that wheat flour also contains gluten. Gluten, mixed with water, forms a strong elastic material that holds the bubbles in place. Many other familiar foods, such as angel food cake and meringues, are solid foams.

A natural solid foam of great practical importance is the bark of the cork oak (*Quercus suber*). Cork is highly anisotropic, which means that it has different properties in different directions. A horizontal slice has the usual solid foam structure, but in a vertical slice, the plant cells are highly elongated (Fig. 3.5). The familiar cork stopper used in the wine

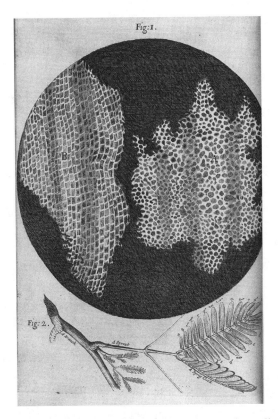

Figure 3.5. A slice of cork showing the different shapes of its cells, depending on whether the cork is cut horizontally or cut vertically. (Robert Hooke, *Micrographia*, 1665)

industry is a cylinder cut with its axis in this vertical direction. As the cork slides into the bottle, the elongated cell walls of the cork will buckle, compressing its diameter by about 30 percent and ensuring a good firm seal. The next time you crack open a bottle of wine, try squishing the cork lengthways versus sideways—one is far easier than the other.

The plant realm is not the only one to produce solid foams. The spittle-bug, or leaf hopper, is a common insect that makes use of liquid foam in the early summer. The nymph of the spittlebug ingests sap from a plant and produces liquid bubbles from its anus. It's a great strategy. The foam dries white, completely hiding the nymph underneath; predators find the taste repellent, and the foam also provides protection from the withering heat of day and the damp chill of night. In time, the nymph emerges as an adult.

Many solid foams have a "high" polymer, such as polystyrene, as their solid component. Such foams are of very low density and are good insulators of both heat and sound. Ear protectors often feature such foam, for the bubbles in the foam absorb sound between 10,000 to 1,000,000 cycles per second, which removes the higher-frequency sounds that we can hear (our ears detect sounds from about 50 to 16,000 cycles per second). And foam has long been used as a way to insulate homes; thick layers of foam prevent heat from escaping through the rafters—and it's cheap.

What's more, solid foams can be good shock absorbers, as NASA discovered. To provide a good night's sleep for astronauts, NASA developed a viscoelastic memory foam. This material is a good choice for a space ship, for it's extremely lightweight, but the temperature- and weight-sensitive foam "flows" around the dozing astronaut, continuously molding to the astronaut's body shape and position in bed, distributing itself and supporting the body weight evenly. It helps both to avoid pressure points and to keep the spine properly aligned, so that the body is less prone to restless sleep. You can now go to one of the many mattress stores and find a viscoelastic foam mattress to help you sleep—it's space-age technology.

Solid foams may help you sleep, but they can also help you walk. Back in the 1960s, Doc Marten's boots first became popular. These shoes sported the famous "air-cushioned sole," which is another example of a solid foam. Today, television commercials urge consumers to go gellin', to get shoes (or inserts) that contain a gel in the sole. Gels, though, are a combination of liquid and solid, close cousins to a solid foam.

The vast majority of solid foams are manufactured, usually by making a liquid foam and then solidifying it. You can freeze the liquid foam, or add suitable chemicals, to make it solid. Engineers can do this for an impressive array of materials, from glassy oxides to ordinary metals, as well as such familiar plastic foams as polyurethane. The final structure of the solid foam is important. It may keep the cell faces of its liquid parent, in which case it is a *closed-cell foam*. Or these may be removed, leaving only the cell borders (thin channels of liquid), which is the case of an *open-cell foam*. In practice, there is not always a clear distinction, for cells may be missing or punctured. This imperfection is desirable when you are baking the solid foams we call bread and cakes, because it prevents them from collapsing when cooled.

Engineers can even make metal foams, and these may help shape the future of cars. To make a metal foam, add a foaming or blowing agent, such as titanium hydride or zirconium hydride, to the powdered metal. Compress the mixture and then heat it up to the temperature where the metal will melt. This process releases gas from the carefully chosen foaming agent, producing a bubbly mixture that, when cooled, forms a lovely closed-cell metal foam (Fig. 3.6). These processes offer another benefit—engineers can heat the mixture inside hollow moulds, and the expanding bubbly liquid will fill the entire area of the moulds, allowing them to make the solid metal foam in whatever shape they wish.

The great potential for metal foams is in the car industry. Because their rheological (flowing) properties make them good energy absorbers, they can greatly reduce the impact of a car crash. Perhaps more important, metal foams retain the strength of the parent metal at a fraction of the weight. About 20 percent of the structural parts of a car could be substituted by metal foams, reducing the car's weight by about 60 kilograms, or 130 pounds. This should reduce fuel consumption significantly.

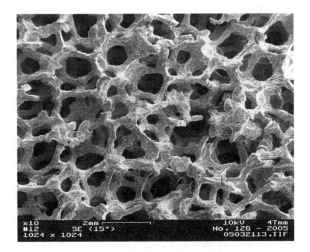

Figure 3.6. Closed-cell metal foam. (SecretDisc/Wikimedia)

The only drawback, currently, is the high cost of producing the foam, but as the price drops, look for these materials to be used widely in the car industry.

Bones

The human skeleton provides us a rigid framework for our muscles, and protects the body's organs (principally the heart and lungs in the bony thoracic cavity, the brain in the skull, and the uterus and bladder in the bony pelvis). In a normal bone, there is an outer shell of hard dense material, and inside this shell is a soft spongy material with a cellular structure called *cancellous bone* (Fig. 3.7). Such a structure greatly reduces the weight of the skeleton, while meeting its primary mechanical requirements. If our bones had been solid, we might never have been able to walk upright. The bones of birds take this principle far beyond the bones of humans.

In children, bone growth is a major concern, and as we grow older, doctors are concerned about the aging of our bones. Bone mass can decrease by age 50, particularly in women. By age 80, a quarter of the bone mass may have been lost. *Osteoporosis* is a disease manifested as a decrease in the density of bones and an increase in their brittleness. Osteo-

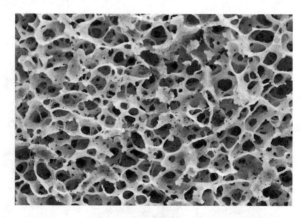

Figure 3.7. The cellular structure of cancellous bone. (Patrick Siemer/ Wikimedia)

porotic bone tissue is less dense than normal bone tissue and appears thin and brittle. A person with osteoporosis is far more susceptible to fractures. Both teenagers and those of advancing years are encouraged to have a diet rich in calcium (milk, green vegetables, and citrus fruits) to help ensure healthy bones. Medicines are available as well. Medical researchers continually seek new materials to be used in place of natural bone. Here, too, metal foams may have a future, for they are strong and lightweight. The research is still in its infancy, but it has already scored a big success. Back in 2005, veterinarians implanted two metal-foam hind legs in Triumph, a Siberian husky.

Mining Foams

As the world's gross domestic product grows, so too does the demand for minerals, such as copper. The price of copper has surged in the past few years, primarily as a result of the huge and rapid growth in the economies of China and India. The severe recent earthquake in Chile, the world's leading producer of copper, has also led to a huge increase in copper prices.

Many millions of tons of mineral ores are treated every year by a process called *flotation*. This process is of great importance economically, and both its use and its application are continually being expanded to treat greater tonnages and to find uses in new areas.

Flotation permits the mining of low-grade ore that would otherwise be uneconomic. The earliest mining venture to use the process for copper ore was undertaken in the Glasdir copper mine, close to Dolgellau in North Wales, by the Elmore family in 1897. Flotation separates the heavy mineral (such as copper), which sinks in the water, from other constituents of the powdered ore (named *gangue*), which float on the water-air surface. The basic flotation process has three steps, which form one circuit. Usually, to extract a reasonable amount of minerals from flotation, several such circuits are necessary.

The first step in the process is to grind the ore into a powder. The mineral and the gangue react differently to water, and these differences can be enhanced by adding specifically tailored chemicals, known as *flotation reagents*. The net result is that, when water is added, the copper hates it but the gangue loves it. Now pass air bubbles through this water-ore mixture, the pulp. Particles of copper flee to the safety of these air bubbles, but to get fine particles of metal to stick to an air bubble is a fine art. There are agitators (like propellers) in the tank that holds the pulp, and these create enough turbulence for air bubbles to form. The agitators also churn up enough copper particles that some will stick to the air bubbles. The bubbles then transport the metal to the surface, forming a foamy froth or, less prosaically, scum (Fig. 3.8). Skim off the scum, which is rich in minerals, and then clean it up to isolate the metal you seek.

Froth flotation, though, is not just for metals. The paper in this book, for example, contains at least 30 percent of recycled materials. Froth flotation is exactly the process by which paper manufacturers take our discarded newspapers and cardboard and strip the inks from them, to make a sheet that is fit for printing.

The world's biggest oil field, in Ghawar, Saudi Arabia, produces more than five million barrels of oil a day. But the extracting of oil isn't easy; it, too, relies on foam technology. When you first drill a hole down into the oil-bearing rock, the excess pressure down below will send the oil shooting up to the surface. But eventually the pressure down there drops, and you will need other ways to extract the remaining oil. One way to do this is to inject water into the geological formation, which forces the oil, which is lighter in weight, to rise to the surface. But this technique leaves

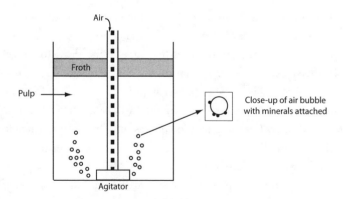

Figure 3.8. Bubbles transported to the surface of a tank carrying minerals, which cling to the bubbles as they rise. The agitator enhances the process. (Greg Nicholl)

about half of the oil still in place, dispersed in droplets throughout the pores of the rock. To extract this residual oil, a foam is pumped through the porous rocks. This foam is rich in surfactants, which allow water and oil to combine, and this fluid can then be pumped to the surface. A tedious, expensive undertaking, but the results usually justify the process.

Foams on the Farm

Cows are amazing creatures; feed them mere grass and they produce milk, in prodigious quantities. But they also produce manure—in equally vast quantities. A fully grown dairy cow yields about 150 pounds of manure in a single day, or about 50,000 pounds per year. In the American Midwest, manure is put to good use: it is placed in so-called lagoons to decompose until it is ready to be sprayed onto the fields as fertilizer. According to the *Wall Street Journal*,* one farmer in Indiana has a problem. His lagoon, containing over 21 million gallons of manure, is producing a nasty foam. Bubbles of methane, generated by the bacteria that feed off the manure, are forming a giant head on the lagoon. Some of the bubbles, more than 20 feet high, are visible on satellite images of the area. Now we don't want these brown bubbles to float off from the top of the lagoon and drift over the neighborhood. After all, methane is a highly flammable gas. The farmer in question, as one might expect, is receiving complaints about the bubbles. But he is a man with a plan—he'll use a small rowboat, wear a gas mask, and pop the bubbles with his trusty Swiss army knife. The worrisome thing: he smokes.

Still, the bubbles may hold the key to the future. Methane, as a fuel, is fairly clean. Those lagoons, therefore, could produce huge amounts of it, which could be burned in power plants to fill some of our energy needs. And this biofuel, as long as we have cows, will never be in short supply.

* Lauren Etter, "Bubbles Boil into Trouble," *Wall Street Journal*, March 25, 2010.

CHAPTER FOUR

Where There's Life, There's Soap

The Science of Surface Tension

The bubble winked at me and said
"You'll miss me, brother, when you're dead"
—*Oliver Herford (1863–1935);*
Toast: The Bubble Winked

A thin steel needle, placed gently on water so that it does not break the surface, floats.* How is this possible? After all, the density of steel is 8 grams per cubic centimeter, whereas that of water is a mere 1 gram per cubic centimeter. The answer? Molecules close to the surface of the water experience an environment quite different from those in the interior of the water. A molecule anywhere in the bulk of the water experiences forces from all directions, as its neighboring molecules tug it toward them. Molecules near the surface of the water, by contrast, experience the tugs from the hemisphere of molecules below it, but none from above, and this differential creates a net force downward. This means that the surface has a tension which, obviously enough, is called *surface tension.*

What we do with great care with a needle on water, nature does with ease. Various insects and other small invertebrates, such as the water spider, can scurry about on the surface of a pond without getting wet.

---

* This chapter is inspired by two famous books: *Soap Bubbles* by C. V. Boys (London: Heinemann, 1965) and *The Science of Soap Films and Soap Bubbles* by Cyril Isenberg (New York: Dover, 1992).

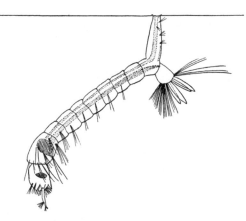

Figure 4.1. A mosquito larva, its breathing tube at the water's surface. (USDA/W. V. King, G. H. Bradley, and T. E. McNeel)

The common gnat and the blood-sucking mosquito, responsible for the transmission of malaria and yellow fever, lay their eggs in stagnant water, and the larvae that hatch out make use of surface tension to hang from the surface of the water when they come up to breathe. They do this by opening flaps at the end of a breathing tube through the tail (Fig. 4.1). If, however, the surface of the water is then covered with a thin layer of oil, which reduces its surface tension, the larvae are unable to support themselves by hanging from the surface and soon perish. By this simple means, areas of land that had hitherto been dangerous or even deadly, such as that within the Panama Canal Zone, have been made habitable.

Soap Bubbles

Soap bubbles have two surfaces: air-to-soap solution and soap solution-to-air. Compare this to an air bubble immersed in water, which has only one surface.

In the nineteenth century, Joseph Antoine Ferdinand Plateau (mentioned in chapter 2) demonstrated something profound. If you dip a wire framework into a bath of soap solution, the shape of the soap film that results *always* has the least possible surface area. Of all the possible ways that a soap film could join, say, two rings together, it chooses the shape that has the smallest surface area, because the smallest surface area requires the least amount of energy. This startling fact, which rapidly attracted the attention of mathematicians, has resulted in a fruitful interaction with experimental scientists.

Plateau devoted much of his days to the study of soap films, despite being completely blind in the latter half of his life. His early research interests were in optics, and in 1829 he performed an experiment in which he exposed his eyes to the sun's rays for 25 seconds. This caused permanent damage to his sight, which gradually deteriorated, so that by 1843, at the age of 42 years, he was *completely* blind. Plateau continued his work in science. He composed a two-volume study, *Statique expérimentale et théorique des liquids soumis aux seules forces moleculaires* (*Experimental and Theoretical Investigations of the Equilibrium Properties of Liquids Resulting from Their Molecular Forces*), which was published in 1873 when he was 71. The book was well received, even though the great James Clerk Maxwell, at Cambridge, was a bit ironic in reviewing it:

> Here, for instance, we have a book, in two volumes, octavo, written by a distinguished man of science, and occupied for the most part with the theory and practice of bubble-blowing. Can the poetry of bubbles survive this?

Figure 4.2 shows that the soap film joining the two rings is a *catenoid*, so called because it looks like a chain, which is *catena* in Latin.

Often, when you observe a soap bubble, you see captivating iridescent patches of color on its surface. White light from the sun, containing all the colors of the rainbow, may reflect back from the outer surface of the film. Then again, light may reflect back from the inner surface of the film. These two sets of reflections interfere with each other, producing the vivid display that we can see, the colors produced depending critically on the thickness of the film. In 2008, Hamid Kellay and his coworkers at the University of Bordeaux put this phenomenon to wonderful use. They created a hemispherical soap bubble 10 centimeters across and heated it from below: the swirling colors that the bubble produced look just like the pattern of hurricanes in the earth's atmosphere; they even created a pattern that mimicked the famous Great Red Spot on Jupiter. The scientists suggest that you can create your own mini-Jupiter

Figure 4.2. Soap film joining two rings. (Photograph courtesy of Colleen Condon)

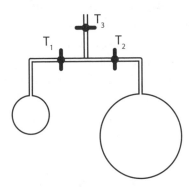

Figure 4.3. The smaller bubble will decrease in size and the larger bubble will expand. (Greg Nicholl)

by using a soap solution and a straw, and heating the bubble to 50°C—at the top end of the temperature range for air emerging from a hair dryer.

The air pressure in a soap bubble lessens as the radius of the bubble increases. As all children know, gentle blowing produces big bubbles, while a sharp burst of air produces tiny bubbles. Take a look at Figure 4.3. What happens if we open taps T_1 and T_2 (T_3 is closed)? Not what many people predict—because the smaller bubble has the greater pressure, air flows from the smaller bubble to the larger bubble. Instead of ending up with two bubbles of the same size, the smaller bubble shrinks while the larger bubble expands.

Coffee Rings

The surface of a liquid at rest is usually curved where it meets a solid surface. Figure 4.4 shows the surface, called a *meniscus*, of each of three liquids in narrow glass tubes. (*Meniscus* comes from a Greek word meaning "crescent," which is an accurate way to describe it.) Suppose you have water in a straight glass (its sides are parallel). The glass is thus a nice vertical surface. There is a point (actually a circle) where the air, the water, and the glass all meet. Draw a line that is tangent to the air-water surface. The angle between this line and the solid surface is called the *angle of contact*. A perfectly flat air/liquid surface, then, would have a contact angle of exactly 90 degrees. The contact angle indicates how much the liquid "likes" the solid. If the liquid wants to spread out over the solid, then the contact angle will be less than 90 degrees and the surface is concave. If the liquid hates the solid, the contact angle will be greater than 90 degrees, the surface of the liquid will be convex, and a droplet of that liquid does not want to spread out on a table made from that solid. For water in a clean glass, the contact angle is close to zero; for paraffin

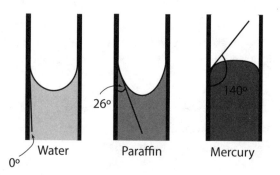

Figure 4.4. The meniscus of three liquids in narrow glass tubes. The contact angle for water is approximately zero. (Greg Nicholl)

in a glass, the angle is about 26 degrees; and for mercury, it is a lofty 140 degrees.

The chemical industry depends on progress in the science of surface phenomena, which affect the wetting, dying, foaming, coalescing, and emulsification processes. Consequently, soap films and soap bubbles are studied extensively in industry and at universities. For example, in recent years biochemists have studied the biological membranes so often present in animal and plant cells. These membranes are composed of lipid molecules similar in structure and behavior to those of soap. By understanding them better, scientists hope they can create tailor-made medicines that will pass through the membranes and into the cell. Such gene therapy could help children with cystic fibrosis or other fatal genetic diseases.

Scientists go to great lengths to explain marvelous natural phenomena—such as coffee rings. Your average droplet of coffee is a mixture of coffee particles and water, the particles perfectly distributed in the water. Spill some of your coffee and later, on the table top, there will be a coffee ring, but its coffee particles will be concentrated on the outside edge of the ring. A group of researchers at the University of Chicago investigated this unexpected finding thoroughly, to try to explain why the particles aren't spread evenly across the entire ring. Where you spill the coffee, its outer edge makes an angle of contact with the table. As the water in the coffee evaporates, you'd expect the coffee itself to shrink, that is, for its outer edge to move back toward the center of the spill. But if the contact point and angle of contact *can't* move, something else has to happen: the bottom layers of the coffee-water mixture start to flow. This brings the coffee particles toward the outer edge and depletes the supply of coffee in the middle of the spill. In other words, the center of the spill will evaporate, leaving a ring of coffee with the grounds concentrated on its outer edge. What started out as a way to explain a curiosity from everyday life may bear rich rewards: printers want ink to hold well on a paper surface so that fine lines won't break up or become blotchy. By knowing how a coffee ring spreads, we may learn how to prevent ink from doing the same.

Soap Molecules

Soap solutions are remarkable for their capacity to form stable bubbles and films. This is because of the surface structure of the soap solution and film, which is a monomolecular (one molecule thick) layer of amphiphilic (from the Greek words meaning "both" and "loving") ions. Each of these ions (electrically charged molecules) has two dissimilar parts. One part is hydrophilic (water-loving): it likes to be surrounded by water. The other part of the ion is hydrophobic (water-hating): it has a dislike for a water environment.

One of the latest developments in materials science is known as Janus spheres, named after the Roman god Janus, who had two faces, the second looking rearward. These are spheres that are given dimples, chemically, so as to resemble microscopic golf balls. The top hemisphere of each sphere is coated with highly water-repellent molecules; the lower hemisphere by highly water-loving molecules. The Janus spheres can then be sprinkled onto an air/water surface where they instantly form a flexible but highly impenetrable barrier. One potential application? Waterproof cosmetics.

Yet again, though, we are outdone by nature. There's a beetle that lives in the harsh climate of the Namib Desert, in the southern part of Africa. It rarely rains in the Namib—less than 10 millimeters fall per year—but often there are incredibly foggy mornings. Parts of the desert beetle's back are covered in a water-hating wax. In the dense fog, the beetle points its head in the direction of the wind. The rough parts of its unwaxed regions form nucleation sites on which water droplets condense out of the fog. Droplets also form on the rougher points of the waxed surface but, when large enough, these will flow downward and join the droplets on the unwaxed region. The result? A large droplet forms rapidly and will roll down—like water off a beetle's back—toward its mouth, so the beetle can enjoy a refreshing (and life-preserving) drink in the bone-dry climate.

The Bubble That Lived

Making soap doesn't require much effort. There's evidence that soap made from mixing animal fat with the ashes from the fire was in use

4,000 years ago. The ashes contain sodium hydroxide, which combines with the fat to form a natural soap consisting of the sodium and potassium salts of fatty acids. These fatty acids can include stearic acid (found in animal fat), but also lauric acid (found in coconut oil), myristic acid (found in nutmeg), palmitic acid (found in palm oil), and oleic acid (found in olive oil and peanut oil). Bars of modern washing soap usually consist of a number of these pure soaps.

Nothing lasts forever, so it's natural that a pure soap film will at some point rupture and a bubble will eventually pop. How long a film lives depends on many things, such as the presence of impurities like dust particles, and whether the film contains too much caustic alkali or fat. (To make a long-lasting soap film, use only distilled water.)

Water evaporates from a bubble's surface, and this process, too, reduces the bubble's life expectancy. The humidity present in the surroundings and air currents affects the evaporation rate, so if you want to produce long-lasting bubbles, choose a still, humid day. Scientists often create bubbles and films in a closed chamber whose atmosphere has a saturated humidity to prevent evaporation, and which is free from shocks, vibrations, air currents, and foreign gases. (You can cheat by adding glycerine, which significantly reduces evaporation and stabilizes the film.) The pure soap film in a controlled environment should last indefinitely, but this is not true for all bubbles. Bubbles contain gas at a pressure greater than the surrounding pressure. Once the film has thinned out significantly, the gas can diffuse right through the bubble wall, and the diameter of the bubble thus decreases with time. This effect is greatest for the smallest bubbles, for these contain gas at the greatest excess pressure.

Sir James Dewar, best known for his invention of the Dewar or Thermos flask, created (apologies to J. K. Rowling) the bubble that lived. Using a controlled environment, he produced a bubble with a diameter of 32 centimeters (13 inches) that remained intact for 108 days, well over three months. (During this period, owing to diffusion, the diameter decreased by a few centimeters.) He also produced a disc of soap film, 19 centimeters (8 inches) in diameter, that he kept for more than three years!

Most ordinary demonstrations don't require bubbles that last three months, or films that last three years. A simple soap solution can be prepared using warm tap water plus about 5 percent of any dish-washing liquid, such as Palmolive. This should produce films that last about 15 seconds (add about 5 percent glycerine to get longer-lasting bubbles, and a splash of food coloring to make it more exciting). Stir the mixture thoroughly before use and remove any bubbles that are formed on the surface of the solution. Some commercial bubble-blowing liquids can produce bubbles that last for a few minutes.

The Soapy Computer

One of the truths that we learn early in life is that the shortest distance between two points is a straight line. But what is the shortest path joining three points, or four, or more? Mathematicians are still working on the general problem of connecting an arbitrary number of points by the shortest possible path.

To see how messy this can get, try to connect four towns that lie at the four corners of a square. To make life easier, the sides of the square are of length 1 so that the diagonal, Pythagoras tells us, is $\sqrt{2}$, which is about 1.41. One option might be a beltway (like the I-495 around Washington, D.C.) or, for Britons, a ring road (such as the M25 around London). In our simple problem, this road would be a circle whose diameter equals the diagonal of the square, $\sqrt{2}$ (Fig. 4.5a). The roadway's total length is then the circumference of a circle, and as keen mathematics students know, this is π multiplied by the diameter. So, the beltway is of length $\pi\sqrt{2}$, which is about 4.44 units.

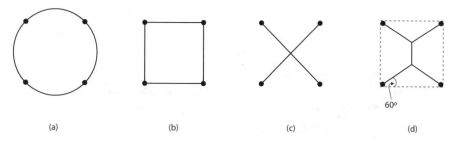

(a)   (b)   (c)   (d)

Figure 4.5. Various ways to connect four towns by different roadway routes. The beltway (a), the square (b), and the cross (c), all of which require more miles of roadway than the last option (d). (Greg Nicholl)

This may seem like a good idea, but if you pave the edges of the square instead (rather than that circular periphery), you have to drive just 4 units (Fig. 4.5b). Another alternative is a giant X made from the two diagonals of the square (Fig. 4.5c). If so, you'd need to pave a paltry $2\sqrt{2}$ units of road, a trifling 2.83 units.

But is this the shortest? To answer this question, we can use a "soap film computer." Remember Joseph Plateau, who told us that a soap film always assumes a *minimum* area? Make a model consisting of two parallel clear Perspex (Plexiglas or Lucite) plates joined by four pins perpendicular to the plates, arranged at the corners of a square. The pins represent, to scale, the four towns. Immerse the model in a bath of soap solution. Take it out and you will see that a soap film has formed between the plates. As per Plateau, the film takes the minimum area, and therefore connects the pins with the least amount of film. The shape of the film indicates the shape the roads should take. The minimum length of road that can link the four towns consists of two three-way intersections. Each intersection consists of three roads meeting at equal angles of 120°. Those eager to have their geometric skills tested can calculate the length of this minimum roadway, which is $1 + \sqrt{3} = 2.73$ units (Fig. 4.5d).

This plan is approximately 4 percent shorter than the value of 2.83 for the usual crossroads system. This may not seem like much of a saving, but the Michigan Department of Transportation recently put the cost of producing a mile of roadway—in a rural area—at $9 million per mile. In an urban area, they estimate a staggering $39 million per mile.

The soap computer can solve problems containing any number of points, or towns, easily. It can determine the minimum length of road, electricity cable, gas pipes, or telephone cables that you need to link towns. Alas, it's a good theory. In practice, roads often deviate from the straight line to take account of geographical constraints such as mountains and lakes, that would increase the cost of construction appreciably. Sociological constraints, such as needing to avoid population centers, also affect the actual route. But it would be wise for planners to be familiar with the more general results concerning the shortest path linking a number of points when attempting to link several towns by roads or pipes. Perhaps air traffic controllers, too, should take a look: what better way to connect the major airports than by routes that yield the least time and the least fuel consumption! Professor Cyril Isenberg, of the University of Kent in England, appeared on national television with his soapy answer to England's road problems: He made two Perspex plates the shape of Great Britain, joined them with pins that were placed where the largest cities are, and dipped them in a soap solution: the giant bubble produced showed a delighted audience how to minimize the miles between the major cities.

Bubble Rafts

Sir Lawrence Bragg was my professor at the Cavendish Laboratory, Cambridge. When he entered the lecture room, my neighbor murmured, "Here we go for another delightful meander through the fertile field of physics." Bragg, who received the Nobel Prize for physics at the tender age of 25, explored how atoms were arranged in crystalline solids. (His father, William, had to wait until he was 53 before receiving the Nobel!) But Sir Lawrence also knew a thing or two about bubbles. He realized that if a large number of hemispherical bubbles of equal size are produced on the surface of a bath of soap solution, using a uniform pressure jet of air, they tend to form a honeycomb lattice of bubbles. Bragg used such rafts of bubbles to show that two-dimensional bubble lattices have many features in common with the crystalline solids he usually studied, and, in his honor, these are named Bragg rafts. These rafts—a fun way to show simple solid-state physics—are now used to study how foams flow.

All of these two-dimensional bubble-raft demonstrations simulate the arrangement of atoms in crystalline materials. Dislocations, lattice defects, grain boundaries, and recrystallization are all phenomena that occur in three-dimensional crystalline materials.

Anti-Bubbles

Sometimes, when soap solution is poured gently from a beaker onto a bath already containing a soap solution, individual drops remain on the surface and move across it just like ball bearings moving on a flat surface. They move away radially from the point of impact. These drops, repelled by the walls of the container but attracted to one another, tend to form groups and coalesce into larger drops. These droplets are unstable and have a lifetime of about 10 seconds. At that point, they suddenly coalesce with the bulk of the soap solution.

If the soap solution is poured rapidly into the bulk solution, it sometimes penetrates deeper into the bulk solution and forms an isolated "bag" of soap solution surrounded by a thin skin of air, though this can be difficult to accomplish. With experience, a "bag" of soap solution with a diameter of up to 20 millimeters can be produced. A "bag" is a spherical globule of soap solution surrounded by a thin shell of air, which in turn is surrounded by soap solution. This is the anti-bubble (Fig. 4.6), the exact opposite of a soap bubble.

An anti-bubble will move, under its own momentum, down into the bulk of the fluid. But because the air shell is less dense than the soap solution, the buoyancy of the surrounding air provides an upthrust that slows the anti-bubble's pace and eventually forces it back to the surface. On reaching the surface, it will either bounce back down into the fluid or come to rest under the surface as a sphere or hemisphere.

The shell of air on the anti-bubble, which is approximately 0.001 millimeter thick, can generate interference colors. These anti-bubbles may

Figure 4.6. The anti-bubble, a spherical globule of soap solution. (Photograph courtesy of Terry Fritz/antibubble.com)

last from seconds to several minutes. They disintegrate suddenly, like the fluid droplets, and produce numerous air bubbles.

Feeling the Tension

Surface tension, it turns out, is the enemy of gravity. A droplet on a surface wants to spread out thinly, under the force of gravity, but the spreading generates a larger surface area. The greater the surface area, the greater the surface tension energy. Eventually, a compromise is reached, and you get a bead of liquid on a surface. (If there were no surface tension, the droplet would spread infinitely!)

The rise of liquid in a thin tube (called *capillary rise*) was studied by Leonardo da Vinci, who showed that the narrower the tube, the greater the rise. This is the principle behind blotting paper for ink, or the use of paper towels to mop up household spills. But a more elegant application of capillary rise is chromatography. If you draw a thick line of black ink on a sheet of absorbent paper, then dip the paper into water, the water travels up the paper, absorbs some of the ink, and continues its traveling. Because different dyes within the black ink travel at different speeds, you can soon see what colors were blended together to make that initial black color. In a chemical laboratory, ink is replaced by the substance you wish to analyze, the paper is replaced by chromatography paper, and the water by a solvent. Invented in the mid-1940s, chromatography was used to explore chlorophyll, which is what gives plants their green color.

Surface tension is a dynamic effect. If you throw a stone into a pond, ripples result. The motion of these tiny waves is dominated by gravity, but also by surface tension. The speed of these departing waves depends on the surface tension, and this information provides another method by which scientists can determine the surface tension between the liquid and the air.

A drop of liquid inserted between two closely spaced horizontal plates will draw them together with a considerable force, provided the angle of contact is less than 90 degrees. If the angle of contact is greater than that, the plates will be repelled with an equally large force. Moreover, the magnitude of the force between the plates increases as the distance between the plates decreases. If you wish to feel the force, put two matching dinner plates close together, nice and snug, and immerse them in soapy water. You'll find that it takes a bit of an effort to separate them. Such "liquid bridges," as they are known, are put to good use by the leaf beetle (*Hemisphaerota cyanea*), which uses about 10,000 minuscule liquid bridges to adhere to the surface of a leaf. And adhere it does: the mighty insect, courtesy of these droplets, can withstand a pull that is one hundred times its own body weight. Inspired by the leaf beetle, Michael Vogel and Paul Steen at Cornell have made a machine that pumps water droplets through a plate that is punctuated by many holes, forming liquid bridges with a plate placed beneath the punctured plate. The bottom plate can support a fairly heavy payload. This finding may one day form the basis of a highly energy-efficient crane.

Another simple home-based experiment demonstrates something similar. Place two blocks of ice close together and allow them to melt slowly. That way, you have two cubes of ice separated by a water layer. The surface tension between them, though, is strong enough to bring the two ice blocks together, and will cause them to re-freeze into a single block.

Two dinner plates are difficult to separate, because the soap between them has an angle of contact less than 90 degrees. Oil, between two metal plates, has an angle of contact greater than 90 degrees, and so *repels* the plates. This is why we use oil as a lubricant in auto engines. The metallic surfaces are repulsed by the oil, which prevents them from touching and sliding over each other—which would cause frictional wear. Water in the engine, though, is a bad idea, for it will have the opposite effect. The metallic surfaces will be drawn together with a considerable force, and the movement of one element relative to the other will produce a large frictional force, resulting in rapid wear and subsequent damage.

When small drops of mercury are dropped onto a clean glass surface, the drops remain spherical, but a large drop is so heavy that it assumes a different shape on contact, and flattens out. Such a drop is said to be *sessile* (meaning "attached to the base"). If you drop some water onto a magazine (newspaper absorbs it too quickly), you can see that the sessile drop has a slightly curved surface, for it acts as a low-quality magnifying glass. Sessile drops have been studied mathematically for well over a century, and their shape can be used to measure the surface tension between the liquid drops and the gas that surrounds them.

Record-Breaking Bubbles

The Science Museum, in London, puts on regular bubble shows. In one show, a demonstrator produces a cylindrical soap film encasing a young child. Those seeking to replicate the feat should use the Science Museum's recipe for soap solution: 7 liters of warm water (in a bucket), 500 milliliters of glycerine (you can get this from a pharmacy), and 500 milliliters of a dish-washing liquid. (These quantities make a huge amount of solution—enough to encase the child—but if all you want to do is blow plain bubbles, use very much less of each ingredient while keeping the proportions the same!) *The Daily Telegraph* reported that on November 24, 2007, a "bubbleologist" packed fifty children inside a soap bubble to break the world record. Sam Heath (Sam Sam the Bubble Man) used a 39-foot (12-meter) wand to create a bubble around the children, who had to be more than 5 feet tall for the record to count.

London's East End holds another bubble world record. Traditionally, the fans of West Ham United sing the song "I'm Forever Blowing Bubbles" whenever their team is winning. On May 16, 1999, the crowd of 23,680 blew bubbles for over a minute, setting a Guinness World Record for simultaneous bubble blowing, thereby celebrating the end of one of their most successful seasons ever.

CHAPTER FIVE

Poetry in the Ocean

Bubbles in the Sea

The world is full of care, much like unto a bubble;
Women and care, and care and women, and women and care and trouble.
—*Epigram that Nathaniel Ward attributed*
to a lady at the court of the Queen of Bohemia
in his book The Simple Cobbler of Aggawam

A Storm at Sea

Few things are as dramatic as a storm at sea. Anyone who lives by the ocean probably spares a thought for sailors whenever storms are brewing. The winds of a sea storm often produce a foam that is stabilized by the constant accumulation of impurities (oil from passing vessels, for example), and the visibly dirty spume that results can be blown far inland.

Below the foam are continuous layers of bubbles. Clouds of bubbles are found at depths down to 20 meters. These bubbles are presumably swept down by turbulent streams arising from the breaking waves. They may persist for 1 to 5 minutes, whereas bubbles in typical sea foam persist for only about 10 to 60 seconds.

Fish interact with bubbles in the sea. Whales, dolphins, and porpoises, all of them mammals, are adept at using bubbles.

Bubble Nets

Humpback whales make bubble nets in which to catch fish. Alone or in groups, humpbacks dive deep, all the while releasing bubbles as they swim in a circle. Each whale dives beneath a shoal of prey and slowly spirals upward, blowing bubbles as it does so. This process creates a hollow-cored cylindrical bubble net, whose interior is relatively bubble-free (Fig. 5.1). The prey congregates in the relatively calm center of the bubble net. The whale then dives beneath the shoal and lunges upward through the bubble net, its mouth open, to gulp down the prey—a tactic known unsurprisingly as *lunge feeding*. Figure 5.2 shows lunge feeding in action. Spiral nets have also been photographed in these circumstances.

But it's probably not the bubbles that cause the prey to gather. It may be that the humpback whales use these bubble nets as a sound trap. When the whales form such nets, they emit loud, trumpeting feeding calls. The wall of the cylinder will partially reflect these noises, creating a relatively quiet interior. Should the prey attempt to leave the trap, it will enter a

Figure 5.1. Aerial view of a bubble net created by a humpback whale. (National Oceanic and Atmospheric Administration)

Figure 5.2. Lunge feeding of humpback whales in action. (Wikimedia)

region where the sound is relatively louder. This may startle the fish, whose natural response is to form schools, and this survival response is transformed by the bubble net into one that aids the predator.

Dolphins have been observed forming a protective wall against sharks by flapping their tails to produce a wall of sound. Recently in New Zealand, a lifeguard was threatened by a shark, but, quite amazingly, nearby dolphins formed one of these protective circles around him, a remarkable event captured on film by BBC Bristol in their *Natural World* program.

Fish-Killing Bubbles

Bubbles can kill fish. In many hydroelectric power plants, fish are often swept down into the turbines by the swift-running water. In some power plants, many fish are killed. But they are not mutilated, suggesting that the fish safely and successfully negotiate the passages between turbine blades. Instead, they succumb to the low pressures and steep pressure gradients caused by bubbles on the turbine blades.

Bubbles Hiding Mines

The importance of mine-sweeping in modern warfare is very great. Since the start of the Cold War, at least fourteen U.S. ships have been damaged by naval mines, some of them sinking. Modern mines are sophisticated, but even old contact mines can be highly dangerous, if not lethal. In 1988, a simple Iranian contact mine costing $1,500 almost sank the *Samuel B. Roberts*, causing nearly $96 million in damage. During the First Gulf War, Iraq laid 1,242 mines, and even though many were nonfunctional or ineffectively laid, three mines seriously damaged two U.S. warships, *Princetown* and *Tripoli*. In very shallow waters, mines can be successfully hidden by the clouds of bubbles generated by waves in the vicinity of the mine.

What Light from Yonder Bubble Breaks?

Sonoluminescence

And the four winds, that had long blown as one,
Shone in my ears the light of sound,
Called in my eyes the sound of light.

—*Dylan Thomas,*
"Love's First Fever to Her Plague"

In 1934, two scientists working at the University of Cologne—Professors Frenzel and Schultes—discovered that bubbles generated in a liquid may produce a faint luminescence that can easily be seen by the naked eye in a dark room. This telltale bluish-white light was dubbed *sonoluminescence*—light from sound.

The source of the light remained a mystery until experiments revealed that the light consisted of flashes, and that these flashes coincided with the frequency of the sound waves. Delicate, precise experiments showed that the flashes occurred when the bubbles collapsed. And when a bubble undergoes successive oscillations, it grows in size, ultimately becoming unstable, and finally collapses, giving out light (Fig. 6.1).

Suppose that the air bubble in water starts off with a diameter of 1 millimeter. After collapse, its diameter might be only a thousandth of a millimeter. This means that its volume will be reduced to a thousandth of a millionth of its original volume, which in turn means that the air inside the bubble is highly compressed. But if you've pumped up a bicycle

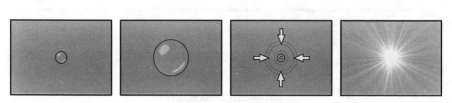

Figure 6.1. A bubble growing in size, then collapsing, at that moment producing sonoluminescence. (Dake/Wikimedia)

tire, you know that it gets hot, too. So the tiny bubble of air becomes enormously heated. Though there is no way to measure its final temperature, we can calculate it to be about 10,000°C. At this temperature, air becomes incandescent and gives off light: this is the hot-spot theory of sonoluminescence.

The Star in the Jar

In 1962, Professor Yosioka in Japan made a major discovery—that a single bubble in water could be trapped and remain stable. (Figure 6.2 shows his simple apparatus.) His results were confirmed, in 1990, by Professor Felipe Gaitan in Mississippi. The method is simple, at least in principle. Fill a round-bottomed glass flask with pure water from which all the gases (such as air) have been removed. Put an oscillating voltage across the vessel, causing the water to vibrate. Thanks to cavitation, this vibration creates a single vapor bubble at the center of the spherical flask. The bubble thus produced is extremely stable and emits a steady glow of bluish light—sometimes for days.

The radius of such a sonoluminescing bubble expands to a maximum and then sharply collapses. The collapse is followed by a series of after bounces, of decreasing amplitude, showing the remarkable elasticity of the bubble. (It behaves much like a rubber ball that bounces repeatedly on the ground, the height achieved decreasing after each bounce.) The whole process then repeats itself in the next sound cycle.

The main advantage of single-bubble sonoluminescence is that the temperature, pressure, and gas content of the bubble can be controlled precisely.

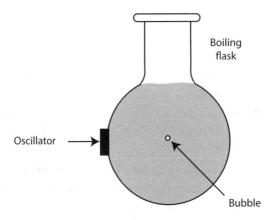

Figure 6.2. An apparatus used to trap a single bubble in water. (Greg Nicholl)

Sonoluminescent Sources

The "star in a jar" is a great venue for studying sonoluminesence, but you can observe it in the world around us. Streams of water, as well as water from hydroelectric power stations, can generate the characteristic blue light. In your home, a water hammer—the annoying clattering of water pipes after a toilet stops flushing, or when a washing machine or dishwasher switches off the water flow—can also produce sonoluminescence.

Animals can create the effect as well. The snapping or pistol shrimp has a giant claw that can be snapped rapidly, producing a thin water jet. The reduced pressure that is left behind allows the tiny air bubbles that exist in seawater to expand rapidly. As the water pressure returns to normal, the bubbles collapse, generating a shock wave strong enough to stun or kill small creatures nearby. The victim is then picked up by a second, normal-sized claw and eaten. The collapse of the bubbles produces weak sonoluminescent light, which some biologists call "shrimpoluminescence." But snapping shrimp live in large colonies (in California's San Diego Bay and around Florida, for example), and the sound generated by the concert of collapsing bubbles can be loud enough to disturb submarine communication.

Liquid Light

In 1675, the French priest and astronomer Jean Picard noticed that when a barometer was carried about in a dark room there was a glow above the mercury contained in it, but could not explain it. In 1709, Francis Hauksbee (the Elder) reported, in his charmingly entitled book *Physico-Mechanical Experiments on Various Subjects. Containing an Account of Several Surprising Phenomena Touching Light and Electricity*, that when mercury was shaken violently in a globe containing air at atmospheric pressure, "particles of light appeared plentifully, about the bigness of pinheads, very vivid, resembling bright twinkling stars." When the air was removed from the same vessel, "the mercury did then appear luminous all round, not as before, like little bright sparks, but as a Continued Circle of light during that motion."

In 1962, Heinrich Kuttruff, of the University of Aachen in Germany, examined this mystery of mercury. He removed the air from a glass tube, partially filled the tube with mercury, and shook the tube backward and forward in the dark, thereby generating weak bluish light. Light appeared as soon as the mercury moved with sufficient speed. With more vigorous shaking, the mercury formed cavities between itself and the glass wall, which rebounded into the mercury with a sharp crack. This generated a second light show: occasional "pin-point" flashes were seen in the cavitation zone.

Mercury is the only metal that exists as a liquid at room temperature. It has a high density (13.6 grams per cubic centimeter; the density of water is a mere 1.00 gram per cubic centimeter). Liquid metals, of which mercury is just one example, are of great practical value; many metals are liquefied in industrial furnaces, for example to help produce the steel for planes, trains, and automobiles. In nuclear submarines, liquid metals are used to help cool down the reactor, thus to keep everyone on board safe.

I became fascinated with liquid metals, and with mercury in particular, and so decided to measure the amount of sonoluminescence from mercury for my Ph.D. thesis at the University of London's Imperial College. The difficulty was clear: mercury is opaque. How can we obtain light from a substance like mercury? The ingenious apparatus that can do this is shown in Figure 6.3. Mercury (or any other liquid metal you choose to study, such as molten tin, copper, or iron) is contained in a small furnace. Sound vibrations are produced electrically and transmitted to the mercury in the furnace. The vibrations generate bubbles in the mercury, and these bubbles produce sonoluminescence. The device, made of glass, that transmits the sound vibrations to the mercury is a velocity transformer. So, sonoluminescent light formed at the bottom of the transformer can travel upward. Waiting for it, at the top, is a photomultiplier tube, one of physics' greatest inventions. The weak light from the mercury enters the photomultiplier tube, which converts it into a stronger electrical signal that is easy to detect and analyze. This allowed

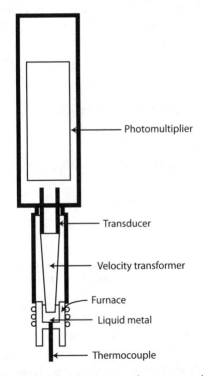

Figure 6.3. The apparatus used to measure the amount of sonoluminescence from mercury, a liquid metal. The thermocouple measures the temperature in the furnace, while the photomultiplier receives the sonoluminescent light emitted by the mercury and amplifies the signal, so that its intensity can be determined. (Greg Nicholl)

me to measure the intensity of the sonoluminescent light produced by the mercury, which turned out to be far greater than that from water.

Light from Breaking Glass

The Italian priest and physicist Giovanni Battista Beccaria (1716–1781) observed something odd. Glass spheres containing air at reduced pressure, when broken, emit light. In 1769, Joseph Priestley, famous for his discovery of oxygen, described the priest's experiments in his book *History of Electricity*, as follows:

> Signor Beccaria observed that hollow glass vessels, of a certain thickness, exhausted of air, gave a light when they were broken in the dark. By a beautiful train of experiments, he found, at length, that the luminous appearance was not occasioned by the breaking of the glass, but by the dashing of the external air against the inside, when it was broke.

Modern equipment gives us a great advantage over the experimentalists of the eighteenth century. A firing pin shatters the glass sphere, a sphere of gas implodes, and a flash of light is emitted. A high-speed cinematograph records the implosion, which lasts for 4 milliseconds. The so-called light divider lets experimenters measure the red and blue contents of the sonoluminescent light.

Sonoluminescence in Space

In 2000, Tom Matula flew his sonoluminescence apparatus on board NASA's parabolic research aircraft (nicknamed "The Vomit Comet," though the NASA authorities prefer "The Weightless Wonder") to determine what effect gravity has on sonoluminescence. Often used for astronaut training, the aircraft flies in parabolic trajectories rather like a roller coaster passing over and between humped stretches of its track. This trajectory exposes the participants and the equipment to bursts (lasting about 25 seconds) of microgravity (near zero gravity) and hypergravity (almost double normal gravity). Matula discovered that sonoluminescence increases by about 20 percent in a weightless environment.

CHAPTER SEVEN

In the Operating Theatre

Medical Bubbles

A little huff, a little blow,
And up the magic bubbles go.
Some large, some small,
Some not lasting long at all.
—*Anonymous*

Cleaning a Wound

There's hardly a bathroom cabinet in the United States that doesn't contain a bottle of hydrogen peroxide. It's easily recognizable, for it comes in an opaque brown bottle. Hydrogen peroxide is a liquid that you pour on minor cuts and abrasions. Ordinary water consists of two hydrogen atoms for every oxygen atom, which chemists describe compactly as H_2O. Hydrogen peroxide has one hydrogen atom per oxygen atom, and its chemical symbol is H_2O_2. In other words, it has "extra" oxygen that it is more than willing to shed. Pour hydrogen peroxide on normal skin and nothing happens. Pour it on a cut and it reacts with an enzyme in the blood and gives off oxygen, copiously. The oxygen foam it produces carries up, and out of, the wound, any dirt and debris that's there, thereby helping the healing process. Its effect is physical, not chemical.

The Bends

What lies beneath the sea? From ancient times, humans have swum below the ocean waves to hunt fish, find pearls, or recover treasure. But in the nineteenth century, the diving helmet was invented, allowing divers to go far deeper than ever before. But the deeper you go, the greater the pressure of the water surrounding you. The pressure of sea water increases by one bar (i.e., one atmospheric pressure) with every 10 meters of depth. Water at 20 meters below the surface thus exerts three times the pressure on a diver that air at sea level does. At great depths the oxygen and nitrogen gases (the natural components of air) that the diver breathes in will dissolve more easily, since they are under greater pressure, and so they spread far more rapidly into the tissues of the body. But if a diver then surfaces too quickly, rapid decompression occurs. Just as droplets of dew form on leaves and plants when the air cannot hold any more water, so gas bubbles form in the diver's tissues and blood, to produce the agonizing pains of the bends, or pain in the joints. To avoid this, divers interrupt their return to the surface at several specified depths, allowing time for the lungs to remove the extra gas accumulated in the blood stream.

Symptoms, including skin itching and mottling, may appear any time within 24 hours after a dive, and they are not pleasant. But the obvious symptom is excruciating joint pain, particularly in the shoulders and knees. Your body contorts to alleviate the pain, hence the name "the bends." The nervous system can also be impaired, causing leg weakness, visual disturbances, or problems with balance, and these are particularly serious. So, too, is a tight, painful feeling across the chest, which may indicate the presence of bubbles in the vessels that supply blood to the heart and lungs.

Any diver with the symptoms described should be placed immediately in a recompression chamber, whose pressure is raised by pumping in air. This extra pressure dissolves the bubbles within the diver's tissues, and the symptoms disappear. Subsequently, the air pressure in the chamber is slowly reduced, allowing the excess gas to escape safely from the blood and lungs.

If treated properly by recompression, most divers with the bends make a full recovery, but in serious untreated cases there may be long-term complications such as partial paralysis. Repeated episodes can lead to degenerative disorders of the bones and joints. A chronic disease of divers who have experienced repeated decompression is a bone disease known as caisson disease (a caisson is a watertight tower that divers work in).

But it's not just divers who can experience the bends. During the construction of the road tunnel under the River Tyne in England, a large group of men worked eight-hour shifts under pressures of about 28 pounds per square inch (about double atmospheric pressure). Out of 5,465 decompressions, 118 (or 2.2 percent) resulted in attacks of the bends. The "sandhogs" who constructed the Queens-Midtown Tunnel that goes under the East River to connect Long Island with downtown New York City suffered more than 300 attacks of the bends. Twenty-five cases involved impairment of the central nervous system, but no sandhogs died because of the bends.

Mountaineers know well that the air is thin at high altitudes. At the top of Mount Everest, an elevation of about 29,000 feet, air pressure is just one-third that at sea level. Climb too quickly and high-altitude sickness, a form of the bends, results. Above about 20,000 feet, there's the risk of breathing too little oxygen, which causes a climber to have less energy, and to become confused and possibly disoriented. Nowadays, climbers sometimes carry portable recompression chambers. Accepting the risks that high-altitude climbing can bring, on May 8, 1978, Italian mountaineer Reinhold Messner and Austrian climber Peter Habeler became the first to conquer Mount Everest without using supplementary oxygen. Two years later, Messner would climb Everest without oxygen again—this time solo and during the monsoon season. Many professional climbers, having attuned their bodies to the rigors, have made the climb since, but some climbers have died in the attempt.

Bubbles in the Blood

The bends are not the only way to get bubbles in the blood. If you have occasion to receive intravenous antibiotics, you'll see that the equipment used to inject the liquids into you is fairly complicated. The reason is simple: large air or gas bubbles in the blood, called embolisms, can be deadly. An air bubble may enter the chambers of the heart, forming an airlock that prevents blood from continuing to circulate.

The heart is a magnificently efficient pump—blood takes about a minute to circulate throughout the human body and return to the heart, flowing at about 5 liters per minute, and the heart can sustain this flow for more than 100 years. But there is turbulent flow inside the human heart, as well as the great pressure changes that this pump produces. Because of cavitation (bubble formation), there may be microbubbles in the cardiac chambers, which can cause damage to heart valves or, at the junctions of arteries, may thicken the walls to cause the life-threatening condition known as arteriosclerosis.

Bubbles in the Bladder

For centuries, people have been suffering from stones in their bladders, or in the vessels entering and leaving the bladder. Surgical removal of kidney stones is called *lithomy*, from the Greek words meaning "stone cutting." In the middle of the nineteenth century, the usual procedure was to enter the bladder from the urethra (the canal by which the urine is discharged from the bladder), and less commonly by open abdominal surgery. Records from London hospitals indicate that one out of seven patients died. Nowadays, there is an alternative.

In lithotripsy, a device known as a *lithotripter* emits thousands of shock waves, focused on kidney stones, at intervals of roughly one per second. These shock waves blast the stone apart into tiny fragments that are small enough (less than 3 millimeters) to exit through the urethra. The risk factor is less than one one-hundredth that of surgical renal stone removal. Treatment is swifter, too: lithotripsy takes only 20 minutes, as opposed to the hospital stay required after a surgical operation. Violent cavitation (bubble formation) occurs during lithotripter operations, and it is this cavitation damage—combined with the forces the shock waves exert—that destroys the stones.

Lithotripters were developed chiefly in Germany, over the last 50 years. There are currently more than ten companies producing them, and shock-wave lithoscopy has become the standard treatment world-wide for stones in the kidneys, bladder, and urethra.

The same technology that underpins kidney stone destruction can make your teeth pearly white. In a trip to the dentist, your teeth will probably be treated by an ultrasonic descaling device. This handheld tool, which uses cavitation (bubbles) to remove calcified plaque from the surfaces of teeth, produces a smoother appearance than manual scaling can achieve.

Designer Bubbles

Doctors now have ultrasound images that help them achieve diagnosis. But as any expectant parent knows, ultrasound pictures of babies tend to be grainy and difficult to read. So, to improve the quality of an ultrasound picture, we'd like to add contrast agents. As the name suggests, these agents yield sharper contrasts in the picture. Other procedures involve similar agents: if you have problems with your gastrointestinal system, you may be given a barium swallow—a drink of barium sulfate that stands out well on an x-ray image—or, if the problems are lower down, a barium enema.

X-rays work well on bones, but not so well on soft tissue. Ultrasound imaging can be used for either, and is safer than x-ray imaging. But how to sharpen ultrasound images? In the past twenty years, an exciting new technology has emerged. Small gas-filled bubbles, each with an artificial elastic shell coating, and less than one one-hundredth of a millimeter in diameter, are introduced into the circulatory system. These are the contrast agents. Their size reflects a deliberate attempt to mimic that of red blood cells. The agents can move freely into capillaries and pass through the pulmonary circulation system. Many contrast agents are now available commercially, including Optison, which contains the gas octafluorocarbon. The shell of the Optison agent is made of human serum albumin.

These encapsulated microbubbles can significantly improve the quality of an ultrasound image. In an ordinary ultrasound image, the myocardium (the middle muscular layer of the heart wall) is invisible. The contrast agent is rapidly injected into a vein. Over several heart cycles, it eventually enters the myocardium, making it visible to ultrasound imaging.

Such contrast agents can measure blood flow and drug delivery. With a little chemistry, the contrast agents can be tailored to attach themselves to specific tissues, such as tumors, and can be coated with the drugs and genes needed to help the diseased tissue. It's strange to think that although ordinary gas bubbles in the bloodstream can kill, designer bubbles have saved thousands of lives.

Epilogue

In this book, I've tried to provide a sketch of the science behind bubbles, droplets, and foams. But our scientific understanding is never complete. Rather, it's a work in progress, and there is always another piece—and then yet another—to add to the jigsaw puzzle. And if the scientific quest itself has no end, neither does the attempt to apply our rapidly mounting knowledge. As we saw in chapter 2, the techniques developed by aerosol scientists to create very fine droplets of liquid in air have been applied to the delivering of antibiotic drugs far down into a patient's lungs, where they can do the most good.

I hope this book conveys some of the excitement of scientific discovery and the fascination of technological innovation. Perhaps young readers may be inspired to begin their own scientific journey, seeking to add more to what we already know about the deceptively simple bubble or to find new applications of it—ones that may save lives or otherwise improve the world around us. I feel pleased to have played a part, myself, in helping to make known some of the principles that lie behind cavitation and sonoluminescence. I wish our understanding of such phenomena were complete—but it never will be.

William Caxton, who first brought the printing press to the English-speaking world, was the first to use the word "bubble" in a printed book. In 1481, he wrote that "The water of those wellis sprynge vp with grete bobles." That's a long way back, but the word "soap" has an even longer history, having descended into English from a word in ancient Hittite for a blend of animal fat and ashes—a substance that was used to dye hair rather than to wash in. I'm left with the feeling that bubbles and soap present challenges to linguists, as well as to scientists and engineers!

Physicists portray bubbles with equations, engineers develop new technologies to exploit them, linguists trace the words to ancient languages, and poets use them as metaphors. But to me, none of these do justice to bubbles. I have seen my children, and then my grandchildren, blow soap bubbles and chase gleefully after them on a hot summer's day. At that moment—for me, as for them—the only description worthy of bubbles was this: simple joy.

atom. The smallest particle of an element that can exist either alone or in combination. See also **hydrogen atom.**

cavitation. A cavity in a liquid often takes the form of a bubble. Cavitation is the name for the development of cavities. We can call it the study of bubbles.

compound. A distinct substance formed by the chemical union of two or more elements in definite proportions, such as sodium chloride (salt).

electron. An elementary particle consisting of a charge of negative electricity. See also **hydrogen atom.**

embolism. The sudden obstruction of a blood vessel by an abnormal particle circulating in the blood.

emulsion. A mixture of two substances (usually liquids) that do not dissolve in each other.

entrainment. One kind of thing scattered through another kind of thing. Here, bubbles trapped in water.

fatty acid. A naturally occurring **compound** usually derived from animal or vegetable fat. Its **molecule** includes a long unbranched chain of carbon **atoms** at one end.

gluten. A mixture of **proteins** found in combination with starch in food grains such as wheat. When gluten is kneaded it becomes elastic, because long strands with cross-links are formed.

helical. Spiral in shape.

hydrocarbon. An organic **compound** containing only carbon and hydrogen and often occurring in petroleum, natural gas, and plastics.

hydrogen atom. An atom consisting of a positively charged nucleus (the **proton**) surrounded by a single rotating negatively charged **electron**. It is thus the simplest of all **atoms**. The electron has a mass of one two-thousandth that of the **proton**.

ion. An **atom** lacking its full complement of **electrons**. In the **hydrogen ion,** lacking its one electron.

ionizing. Producing ions.

mass. A measurement of the amount of material a body contains and causes it to have weight in a gravitational field.

meniscus. The curved upper surface of a column of liquid. See Figure 4.4.

molecule. The smallest particle of a substance that retains all the properties of the substance and is composed of one or more **atoms.**

monomer. A chemical **compound** that can undergo **polymerization.**

myocardium. The heart muscle.

nebulizer. A device that converts a liquid (for example water) to a fine spray.

neutron. A particle having the same mass as the **proton** of an **atom** but no electrical charge. See also **hydrogen atom.**

oleic acid. A fatty acid found in various animal and vegetable sources (for example, grape seed oil).

palmitic acid. One of the most common saturated fatty acids. It is the major component of oil from palm trees, often added to processed foods.

photomultiplier tube. An electronic device for measuring very weak light.

plankton. The passively floating or weakly swimming usually minute animal and plant life of a body of water.

polymer. A substance composed of large molecules; specifically, a chemical compound or mixture of compounds consisting essentially of repeating structural units.

polystyrene. A polymer made from the **monomer styrene** (a liquid **hydrocarbon** manufactured from petroleum).

polyurethane. A **polymer** consisting of a chain of organic units joined by **urethane** links. Polyurethanes exhibit a wide range of stiffness, hardness, and density.

protein. One of the extremely complex **molecules** found in humans, other animals, and plants.

proton. A positively electrically charged elementary particle that along with **neutrons** is a constituent of all atomic nuclei. See also **hydrogen atom.**

refracted. Deviated, as light passing through a prism.

saturated fat. A fatty substance such as animal fat that is unable to absorb additional fatty material.

serum albumin. A blood protein that normally constitutes more than half of the protein in blood serum. See also **protein.**

sodium bicarbonate. A white crystalline solid. Baking soda is sodium bicarbonate.

sonoluminescence. Light produced under certain conditions by a bubble receiving a sound wave.

styrene. A liquid unsaturated **hydrocarbon** used chiefly in making synthetic rubber, resins, and plastics.

sulfur dioxide. A choking gas formed during the combustion of sulfur **compounds**, as for example in a coal or tire fire.

sulfuric acid. A strong corrosive mineral acid; a constituent of acid rain.

surfactant. A surface-active agent, such as a detergent used to separate grease from water.

synovial fluid. The fluid that fills a joint, such as a knee or finger.

ultrasound. Sound whose frequency is above the upper limit of human audibility.

uranium. A silver-gray metallic chemical element that is weakly radioactive and 70 percent more dense than lead. It is commercially extracted from uranium-bearing minerals such as uraninite.

urethane. A crystalline **compound** used especially as a solvent, and for some medicinal purposes.

zirconium hydride. A chemical compound of zirconium and hydrogen. It is a flammable gray-black powder.

Boys, C. V. *Soap Bubbles: Their Colours and the Forces Which Mould Them.* London: Heinemann, 1965.

Gardner, Robert. *Experiments with Bubbles.* Berkeley Heights, N.J.: Enslow, 1995.

Gleick, James. *Chaos: Making a New Science.* New York: Viking, 1987.

Hart-Davis, Adam. *Why Does a Ball Bounce?* Buffalo: Firefly Books, 2005.

Hauksbee, Francis. *Physico-Mechanical Experiments on Various Subjects. Containing an Account of Several Surprising Phenomena Touching Light and Electricity.* London: Printed by R. Brugis, for the author, 1709.

Isenberg, Cyril. *The Science of Soap Films and Soap Bubbles.* New York: Dover, 1992.

Kepler, Johannes. *De Nive Hexangula* (*The Six-Cornered Snowflake,* 1611). Oxford: Clarendon, 1966.

Minnaert, M. G. J. *Light and Color in the Outdoors.* New York: Springer-Verlag, 1993.

Plateau, Joseph Antoine Ferdinand. *Statique expérimentale et théorique des liquides soumis aux seules forces moleculaires* (*Experimental and Theoretical Investigations of the Equilibrium Properties of Liquids Resulting from Their Molecular Forces*). Paris: Gauthier-Villars, 1873.

Priestley, Joseph. *The History and Present State of Electricity, with Original Experiments.* London: Printed for J. Dodsley, J. Johnson, and T. Cadell, 1767.

Sitwell, Osbert. *Escape with Me.* London: Macmillan, 1939.

Sutcliffe, James Frederick. *Plants and Water.* New York: St. Martin's Press, 1968.

Trevena, D. H. *The Liquid Phase.* New York: Springer-Verlag, 1975.

———. *Cavitation and Tension in Liquids.* Bristol: Adam Hilger, 1987.

Ward, Alan. *Experimenting with Surface Tension and Bubbles.* Chicago: Trafalgar Square, 1986.

Weaire, Denis, and Stefan Hutzler. *The Physics of Foams*. New York: Oxford University Press, 1999.

Whymper, Edward. *Scrambles amongst the Alps*. London: J. Murray, 1871.

Young, F. Ronald. *Cavitation*. London: Imperial College Press, 1989.

———. *Sonoluminescence*. Boca Raton, Fla.: CRC Press, 2005.